D0344656

# THE YOUNGEST SCIENCE

*Also by Lewis Thomas*

The Lives of a Cell

The Medusa and the Snail

*This book is published as part of an*
*Alfred P. Sloan Foundation Program*

# THE YOUNGEST SCIENCE

## *Notes of a Medicine-Watcher*

LEWIS THOMAS

THE VIKING PRESS
NEW YORK

First published in 1983 by The Viking Press
40 West 23rd Street, New York, N.Y. 10010
Published simultaneously in Canada by
Penguin Books Canada Limited

*Library of Congress Cataloging in Publication Data*
Thomas, Lewis, 1913–
The youngest science.
1. Thomas, Lewis, 1913–
2. Physicians—United States—Biography.
I. Title.
R154.T48A36 1983 610'.92'4 [B] 82-50736
ISBN 0-670-79533-X

"Millennium" was originally published in *The Atlantic Monthly.* Copyright 1941 by Lewis Thomas. Copyright renewed © 1969 by Lewis Thomas.
"Allen Street" was originally published in the *Aesculapiad,* Harvard Medical School, 1937.

Portions of this book were previously published in *McCall's* and *Self*

Printed in the United States of America
Set in Linotron Goudy
Designed by Ann Gold

*To Beryl, Abigail, Judith, Eliza,*
*best friends*

THE ALFRED P. SLOAN FOUNDATION SERIES

Disturbing the Universe
*by Freeman Dyson*

Advice to a Young Scientist
*by Peter Medawar*

The Youngest Science: Notes of a Medicine-Watcher
*by Lewis Thomas*

FORTHCOMING:

Haphazard Reality: Half a Century of Science
*by Hendrik B. G. Casimir*

In Search of Mind: Essays in Autobiography
*by Jerome S. Bruner*

# *PREFACE TO THE SERIES*

The Alfred P. Sloan Foundation has for many years included in its areas of interest the encouragement of a public understanding of science. It is an area in which it is most difficult to spend money effectively. Science in this century has become a complex endeavor. Scientific statements are embedded in a context that may look back over as many as four centuries of cunning experiment and elaborate theory; they are as likely as not to be expressible only in the language of advanced mathematics. The goal of a general public understanding of science, which may have been reasonable a hundred years ago, is perhaps by now chimerical.

Yet an understanding of the scientific enterprise, as distinct from the data and concepts and theories of science itself, is certainly within the grasp of us all. It is, after all, an enterprise conducted by men and women who might be our neighbors, going to and from their workplaces day by day, stimulated by hopes and purposes that are common to all of us, rewarded as most of us are by occasional successes and distressed by occasional setbacks. It is an enterprise with its own rules and customs, but an understanding of that enterprise is accessible to any of us, for it is quintessentially human. And an understanding of the enterprise inevitably brings with it some insight into the nature of its products.

Accordingly, the Sloan Foundation has set out to encourage a representative selection of accomplished and articulate scientists to set down their own accounts of their lives in science. The form those accounts will take has been left in each instance to the author: one may choose an autobiographical approach, another may produce a coherent series of essays, a third may tell the tale of a scientific community of which he was a member. Each author is a man or woman of outstanding accomplishment in his or her field. The word "science" is not construed narrowly: it includes such disciplines as economics and anthropology as much as it includes physics and chemistry and biology.

The Foundation's role has been to organize the program and to provide the financial support necessary to bring manuscripts to completion. The Foundation wishes to express its appreciation of the great and continuing contribution made to the program by its Advisory Committee chaired by Dr. Robert Sinsheimer, Chancellor of the University of California–Santa Cruz, and comprising Dr. Howard H. Hiatt, Dean of the Harvard School of Public Health; Dr. Mark Kac, Professor of Mathematics at Rockefeller University; Daniel J. Kevles, Professor of History at the California Institute of Technology; Robert K. Merton, University Professor, Columbia University; Dr. George A. Miller, Professor of

Experimental Psychology at Princeton University; for the Foundation, Arthur L. Singer, Jr., and Stephen White; for Harper & Row, Edward Burlingame.

—ALBERT REES
President, Alfred P. Sloan Foundation

# CONTENTS

# THE YOUNGEST SCIENCE

# I
# AMITY STREET

I have always had a bad memory, as far back as I can remember. It isn't so much that I forget things outright, I forget where I stored them. I need reminders, and when the reminders change, as most of them have changed from my childhood, there goes my memory as well.

The town I was born and raised in disappeared. The only trace left behind to mark the location of the old clapboard house we lived in is the Long Island Rail Road, which still penetrates and crosses the town through a deep ditch, and somewhere alongside that ditch, behind a cement wall, is the back yard of my family's house. All the rest is gone. The yard is now covered by an immense apartment house. The whole block, and the other blocks around where our neighbors' clapboard houses and back-yard gardens were, are covered by apartment houses, all built fixed to each other as though they were a single syncytial structure. The trees, mostly maples and elms, are gone. The church my family went to, most Sunday mornings, is still there, looking old and beat-up, with a sign in the front indicating that it is no longer the Dutch Reformed Church and now is Korean Protestant. I drove down this block, darkened now like a tunnel by the apartment buildings set close to the curb on each side, and saw nothing to remind me of any part of my life.

Lacking landmarks, I cannot be sure that the snatches of memory still lodged in my brain have any reliability at all; I could have made them up, or they could be the memories of dreams. I do dream about Flushing from time to time, finding myself on a bicycle on Boerum Avenue between Amity Street and Madison Street (all of these street names are gone, replaced now by numbers), and there is the town garbage wagon, horse-drawn, driven by a wild-eyed and red-haired youngish man named Crazy Willie, racing along the block on his high seat, talking to himself. I'm pretty sure of that memory; there was such a garbage wagon, driven by Crazy Willie, but why do I still have the image taped in my temporal lobe ready for replaying so many late nights, and so little else? I remember, now that I think of it, the late Sunday afternoon when the Lawnmower family arrived for a visit, friends of my family from somewhere far away, Ohio maybe, whose name turned out years later to be Lorrimer. It must have been around the same time that I discovered what the maid told me was copper beneath the sandbox, great soggy sheets of friable copper, enough to secure the family's fortune, which I already knew needed securing, and then, a couple of years later, I lost the fortune on learning that she must have been saying carpet. There was a huge cherry tree at the back of the yard, close to the cement fence, and something went wrong with it, death I suppose, it was cut down and chopped up there in the yard, and what remains stored in my brain now, sixty years later, is the marvelous smell of that wood, the smell of the whole earth itself, all over the yard and, for a few days until it was carted away, in all the rooms of our house.

My earliest clear memory of my mother is her tall figure standing alone in the center of the lawn behind the house, looking down at the grass, turning in a slow circle, scanning the ground. From the time of my earliest childhood I knew this to be a mild signal of trouble for my mother, trouble for the family. Sometimes she stood there for only a few moments, sometimes for as long as five minutes. Then, in the quickest of movements, she would reach down to pluck the four-leaf clover she was hunting and come back to the house. If I was there on the back porch, watching as she came, she would laugh at me and say, always the same sentence, "The Lord will provide."

So far as I know, this was her only superstition, or anyway, the only one she ever acted on. And it was always used for the same purpose, which was to get my father's patients to pay their bills.

Very few of the patients paid promptly, and a good many never paid at all. Some sent in small checks, once every few months. A few remarkable and probably well-off patients paid immediately, the whole bill at once, and when this happened my father came upstairs after office hours greatly cheered.

There was never an end to worrying about money, although nobody talked much about it. The family took it for granted that my father had to worry about his income at the end of every month, and we knew that he was absolutely determined to pay all his bills on the first of each month, without fail. He believed that being in debt was the worst of fates, and he paid everyone—the grocer and butcher, the coal man, taxes, and the instrument and drug houses that supplied his office—as soon as he could after the bills arrived, depending on how much cash he had in the bank. But it was not the style of the time to pay the doctor quickly.

These were the years everyone thinks of as the good times for the country, the ten years before the Great Depression. The town was prosperous, but the practice of medicine was accepted to be a chancy way to make a living, and nobody expected a doctor to get rich, least of all the doctors themselves. In the town where I grew up, there were two or three physicians whose families seemed rich, but the money was old family money, not income from practice; the rest of my father's colleagues lived from month to month on whatever cash their patients provided and did a lot of their work free, not that they wanted to or felt any conscious sense of charity, but because that was the way it was.

My father kept his own books, in a desk calendar that recorded in his fine Spencerian handwriting the names of the patients he had seen each day, each name followed by the amount he charged, and that number followed by the amount received. It was the last column that mattered. My mother kept a careful eye on those numbers, and it was always toward the end of the month that she went back to the lawn to find her four-leaf clovers.

I'll never understand how she did it. As I grew older, seven or eight years old, I liked to go along while she sought the family fortune, to help out if I could, but I never spotted a single one, even though my eyes were a lot closer to the ground. We would stand side by side, and I would try to scan the same patches of lawn, staring hard, but even as she swooped down to pick one, I was never able to see it until it was in her fingers.

Much later, when I was a fourth-year medical student at Harvard, I learned more objectively some of the facts of medical economics. The yearbook for the class of 1937 was edited by Albert Coons, my closest friend in the class, and I was invited to be on the editorial staff because I'd written a rather long and disrespectful poem about medicine and death, called "Allen Street." Coons prepared a questionnaire for the book and sent it out late in 1936, to the Harvard graduates from the years 1927, 1917, and 1907. The questions dealt mainly with the kinds of internship and residency training experiences most highly regarded by Harvard doctors ten, twenty, and thirty years out of school, but there were also a couple of lines asking, delicately and promising anonymity, for the respondent's estimated income for 1937, and then a generous empty space at the bottom of the page requesting comments in general, advice to the class of 1937.

Surprisingly, 60 percent of the questionnaires were filled in and returned, and they made interesting reading for Coons and me and all our classmates. Most of the papers neglected the business about postgraduate training and concentrated on the money questions. The average income of the ten-year graduates was around $3500; $7500 for the twenty-year people. One man, a urologist, reported an income of $50,000, but he was an anomaly; all the rest made, by the standards of 1937, respectable but very modest sums of money.

The space at the bottom of the page had comments on this matter, mostly giving the same sort of advice: medicine is the best of professions was the general drift, but not a good way to make money. If you could manage to do so, you should marry a rich wife.

It was very hard work, being a doctor. All the men (there were only men in those Harvard classes) had a line or two about the work: long hours, no time off, brief holidays. Prepare to work very hard was their advice to the class of 1937, and don't expect to be prosperous.

Watching my father's work was the most everyday part of my childhood. He had his office at home, like all the doctors in Flushing. The house was a large Victorian structure with a waiting room and office in the ground-floor area that would have been the parlor and drawing room for other houses of the period. My family had their sitting room on the second floor, but the dining room was downstairs, a door away from the patients' waiting room, so we grew up eating more quietly and quickly than most families.

In the best of times, right up until the start of the Depression, we had a live-in maid who had her room on the third floor and a laundress who worked in the basement; then a part-time maid during the first years of the Depression; finally nobody. My mother always did the cooking, even when there was a maid; later, she did all the cleaning and everything else in the house, and in her free time she worked the garden around the edges of the back yard. We had had a gardener once, I remember, in the early 1920s, an Italian named Jimmy who came up from Grove Street. Jimmy and my mother would discuss the progress of the garden every day, he in rapid torrents of passionate, arm-waving Italian, she in slow, careful, but firmly put English, and they got along fine. Later on, in the Depression years, she gardened the whole place herself, while the children mowed the lawn.

Two streets scared the children away by their strangeness: Grove Street, just beneath the Long Island Rail Road Station, where the Italians lived, several dozen families, all poor, all speaking Italian at home and broken English elsewhere; and Lincoln Street, where the black people lived. Lincoln Street was not a ghetto, it was right in the center of Flushing, the best part of town, but two blocks of Lincoln Street were entirely black. I used to wonder how that happened, why all the Negroes lived together on those two blocks, but it was never explained; it had always been like that.

Memorial Day and the Fourth of July were the town's major events. Both involved parades, the first down Northern Boulevard to Town Hall and the Civil War Monument, where a Boy Scout was required to recite Lincoln's Gettysburg Address (one year I had to do it), the second up from Main Street along Sanford Avenue. The people lined the streets waiting for the open cars containing the Civil War veterans, old, grizzled, confused-looking men in their eighties, wearing Union uniforms, then the World War I veterans (it was called the Great War then, nobody thought of giving it a number), young and fresh, in khaki uniforms and puttees. Brass bands, flags, the Masons and the Knights of Columbus, the village police and firemen, streams of children in Boy Scout, Girl Scout, Girl Pioneer, Sea Scout uniforms, and smaller children from the parochial school in everyday clothes, faces flushed pink with pleasure.

The two most important people in town, known and respected by everyone, were Miss Guy, who taught first grade at P.S. 20, and Mr. Pierce, the principal. Miss Guy was the great eminence, having taught several generations of Flushing people. Mr. Pierce owed his social distinction to the loftiness of his position alone, having arrived from out of town only ten years earlier.

All the children in Flushing were juvenile delinquents. We roamed the town in the evening, ringing doorbells and running around the side of the house to hide, scrawling on the sidewalks with colored chalk, practicing for Halloween, when we turned into vandals outright, breaking windows, throwing garbage cans into front yards, twisting the street signs to point in the wrong direction. We shoplifted at Woolworth's, broke open the nickel-candy machines fixed on the backs of the seats in the Janice Cinema, bought Piedmont cigarettes and smoked them sitting on the curb on Main Street, at the age of ten. A bad lot.

In the time of my childhood, nothing but the worst was expected of children. We were expected to be bad, there was no appealing to our better selves because it was assumed that we had no better selves. Therefore, to be contrary, as is the habit of children, we turned out rather well.

My father never had an office nurse or a secretary. The doorbell was answered by my mother or by whatever child was near at hand, or by my father if he was not involved with a patient. The office hours were one to two in the afternoon and seven to eight in the evening. I remember those numbers the way I remember old songs, from hearing my mother answering the telephone and, over and over again, repeating those hours to the callers: there was a comforting cadence in her voice, and it sounded like a song—one to two in the after*noon,* seven to eight in the *eve*ning.

The waiting room began to fill up an hour before the official office hours, and on busy days some of the patients had to wait in their cars outside or stand on the front porch.

Most days, my father saw ten patients in each hour; I suppose half of these were new patients, the other half people coming back to be checked from earlier visits.

Except for the office hours and quick meals, my father spent his hours on the road. In the early morning he made rounds at the local hospital, where, as chief of surgery, he would see the patients in the surgical wards as well as his own private patients. Later in the morning, and through the afternoon, he made his house calls. In his first years of practice, when he and my mother moved out from New York City to Flushing, which they picked because it was a small country town with good trees and gardens but with the city still accessible by train, he had a bicycle, then a year later a horse and buggy, each of which he detested. A year or so before I was born, he had prospered enough to buy an automobile. First it was a Maxwell, which broke down a lot and kept him in a continual temper, then a snub-nosed Franklin sedan, finally a quite expensive Franklin coupé with a "modern" conventional front.

He spent the major part of his life in these cars, driving to the hospital and then around Flushing and through the neighboring towns, seeing one patient after another. He came home around nine or ten most evenings.

But it was at night, long after the family had gone to sleep, that my father's hardest work began. The telephone started ringing after midnight. I could hear it from my bedroom down the hall, and I could hear his voice, tired and muffled by sleep, asking for details, and then I could hear him hang up the phone in the dark; usually he would swear "Damnation," sometimes he was distressed enough to use flat-out "Damn it," or worse, "Damn"; rarely did I hear him say, in total fury, "God damn it." Then I could hear him heave out of bed, the sounds of dressing, lights on in the hall, and then his steps down the back stairs, out in the yard and into the car, and off on a house call. This happened every night at least once, sometimes three or four times.

I never learned, listening in the dark, what the calls were about. They always sounded urgent, and sometimes there were long conversations in which I could hear my father giving advice and saying he'd be in the next morning. More often he spoke briefly and then hung up and dressed. Some were for the delivery of babies. I remember that because of my mother's voice answering the phone even later at night, when he was off on his calls, saying that the doctor was out on a "confinement." But it was not all babies. Some were calls from the hospital, emergencies turning up late at night. Some were new patients in their homes, frightened by one or another sudden illness. Some were people dying in their beds, or already dead in their beds. My father must have been called out for patients who were dying or dead a great many of his late nights.

Twenty years later, when I was on the faculty at Tulane Medical School and totally involved in the science of medicine, I had another close look at this side of doctoring. I had been asked to come to the annual meeting of a county medical society in the center of Mississippi, to deliver an address on antibiotics. The meeting was at the local hotel, and my host was the newly elected president of the society, a general practitioner in his forties, a successful physician whose career was to be capped that evening, after the banquet, by his inauguration; to be the president of the county medical society was a major honor in that part of the world. During the dinner he was called to the telephone and came back to the head table a few minutes later to apologize; he had an emergency call to make. The dinner progressed, the ceremony of his induction as president was conducted awkwardly in his absence, I made my speech, the evening ended, and just as the people were going out the door he reappeared, looking harassed and tired. I asked him what the call had been. It was an old woman, he said, a patient he'd looked after for years; early that evening she had died, that was the telephone call. He knew the family was in distress and needed him, he said, so he had to go. He was sorry to have missed the evening, he had looked forward to it all year, but some things can't be helped, he said.

This was in the early 1950s, when medicine was turning into a science, but the old art was still in place.

# 2
# House Calls

My father took me along on house calls whenever I was around the house, all through my childhood. He liked company, and I liked watching him and listening to him. This must have started when I was five years old, for I remember riding in the front seat from one house to another, and back and forth from the hospital, when my father and many of the people on the streets were wearing gauze masks; it was the 1918 influenza epidemic.

One of the frequent calls which I found fascinating was at a big house on Sanford Avenue; he never parked the car in front of this house, but usually left it, and me, a block away around the corner. Later, he explained that the patient was a prominent Christian Scientist, a pillar of that church. He could perfectly well have parked in front if there had been a clearer understanding all around of what he was up to, for it was, in its way, faith healing.

I took the greatest interest in his doctor's bag, a miniature black suitcase, fitted inside to hold his stethoscope and various glass bottles and ampules, syringes and needles, and a small metal case for instruments. It smelled of Lysol and ether. All he had in the bag was a handful of things. Morphine was the most important, and the only really indispensable drug in the whole pharmacopoeia. Digitalis was next in value. Insulin had arrived by the time he had been practicing for twenty years, and he had it. Adrenalin was there, in small glass ampules, in case he ran into a case of anaphylactic shock; he never did. As he drove his rounds, he talked about the patients he was seeing.

I'm quite sure my father always hoped I would want to become a doctor, and that must have been part of the reason for taking me along on his visits. But the general drift of his conversation was intended to make clear to me, early on, the aspect of medicine that troubled him most all through his professional life; there were so many people needing help, and so little that he could do for any of them. It was necessary for him to be available, and to make all these calls at their homes, but I was not to have the idea that he could do anything much to change the course of their illnesses. It was important to my father that I understand this; it was a central feature of the profession, and a doctor should not only be prepared for it but be even more prepared to be honest with himself about it.

It was not always easy to be honest, he said. One of his first patients, who had come to see him in his new office when he was an unknown in town, was a man complaining of grossly bloody urine. My father examined him at length, took a sample of the flawed urine, did a few other tests, and found himself without a diagnosis. To buy time enough to read up on the matter, he gave the patient a bottle of Blaud's pills, a popular iron remedy for anemia at the time, and told him to come back to the office in four days. The patient returned on the appointed day jubilant, carrying a flask of crystal-clear urine, totally cured. In the following months my father discovered that his reputation had been made by this therapeutic triumph. The word was out, all over town, that that new doctor, Thomas, had gifts beyond his own knowledge—this last because of my father's outraged protests that his Blaud's pills could have had nothing whatever to do with recovery from bloody urine. The man had probably passed a silent kidney stone and that was all there was to it, said my father. But he had already gained the reputation of a healer, and it grew through all the years of his practice, and there was nothing he could do about it.

Even now, twenty-five years after his death, I meet people from time to time who lived once in Flushing, or whose parents lived there, and I hear the same anecdotes about his abilities: children with meningitis or rheumatic fever whose lives had been saved by him, patients with pneumonia who had recovered under his care, even people with incurable endocarditis, overwhelming typhoid fever, peritonitis, what-all.

But the same stories are told about any good, hardworking general practitioner of that day. Patients do get better, some of them anyway, from even the worst diseases; there are very few illnesses, like rabies, that kill all comers. Most of them tend to kill some patients and spare others, and if you are one of the lucky ones and have also had at hand a steady, knowledgeable doctor, you become convinced that the doctor saved you. My father's early instructions to me, sitting in the front of his car on his rounds, were that I should be careful not to believe this of myself if I became a doctor.

Nevertheless, despite his skepticism, he carried his prescription pad everywhere and wrote voluminous prescriptions for all his patients. These were fantastic formulations, containing five or six different vegetable ingredients, each one requiring careful measuring and weighing by the druggist, who pounded the powder, dissolved it in alcohol, and bottled it with a label giving only the patient's name, the date, and the instructions about dosage. The contents were a deep mystery, and intended to be a mystery. The prescriptions were always written in Latin, to heighten the mystery. The purpose of this kind of therapy was essentially reassurance. A skilled, experienced physician might have dozens of different formulations in his memory, ready for writing out in flawless detail at a moment's notice, but all he could have predicted about them with any certainty were the variations in the degree of bitterness of taste, the color, the smell, and the likely effects of the concentrations of alcohol used as solvent. They were placebos, and they had been the principal mainstay of medicine, the sole technology, for so long a time—millennia—that they had the incantatory power of religious ritual. My father had little faith in the effectiveness of any of them, but he used them daily in his practice. They were expected by his patients; a doctor who did not provide such prescriptions would soon have no practice at all; they did no harm, so far as he could see; if nothing else, they gave the patient something to do while the illness, whatever, was working its way through its appointed course.

*The United States Pharmacopoeia*, an enormous book, big as the family Bible, stood on a bookshelf in my father's office, along with scores of textbooks and monographs on medicine and surgery. The ingredients that went into the prescriptions, and the recipes for their compounding and administration, were contained in the *Pharmacopoeia*. There was no mistaking the earnestness of that volume; it was a thousand pages of true belief: this set of ingredients was useful in pulmonary tuberculosis, that one in "acute indigestion" (the term then used for what later turned out to be coronary thrombosis), another in neurasthenia (weak nerves; almost all patients had weak nerves, one time or another), and so on, down through the known catalogue of human ailments. There was a different prescription for every circumstance, often three or four. The most popular and widely used ones were the "tonics," good for bucking up the spirits; these contained the headiest concentrations of alcohol. Opium had been the prime ingredient in the prescriptions of the nineteenth century, edited out when it was realized that great numbers of elderly people, especially "nervous" women, were sitting in their rocking chairs, addicted beyond recall.

The tradition still held when I was a medical student at Harvard. In the outpatient department of the Boston City Hospital, through which hundreds of patients filed each day for renewal of their medications, each doctor's desk had a drawerful of prescriptions already printed out to save time, needing only the doctor's signature. The most popular one, used for patients with chronic, obscure complaints, was *Elixir of I, Q, and S*, iron, quinine, and strychnine, each ingredient present in tiny amounts, dissolved in the equivalent of bourbon.

Medicine was subject to recurrent fads in therapy throughout my father's career. Long before his time, homeopathy emerged and still had many devout practitioners during his early years; this complex theory, involving what was believed to be the therapeutic value of "like versus like," and the administration of minuscule quantities of drugs that imitated the symptoms of the illness in question, took hold in the mid-nineteenth century in reaction against the powerfully toxic drugs then in common use—mercury, arsenic, bismuth, strychnine, aconite, and the like. Patients given the homeopathic drugs felt better and had a better chance of surviving, about the same as they would have had without treatment, and the theory swept the field for many decades.

A new theory, attributing all human disease to the absorption of toxins from the lower intestinal tract, achieved high fashion in the first decade of this century. "Autointoxication" became the fundamental disorder to be overcome by treatment, and the strongest measures were introduced to empty the large bowel and keep it empty. Cathartics, ingenious variations of the enema, and other devices for stimulating peristalsis took over medical therapy. My father, under persuasion by a detail man from one of the medical supply houses, purchased one of these in 1912, a round lead object the size of a bowling ball, encased in leather. This was to be loaned to the patient, who was instructed to lie flat in bed several times daily and roll it clockwise around the abdomen, following the course of the colon. My father tried it for a short while on a few patients, with discouraging results, and one day placed it atop a cigar box which he had equipped with wheels and a long string, and presented it to my eldest sister, who tugged it with pleasure around the corner to a neighbor's house. That was the last he saw of the ball until twelve years later, when the local newspaper announced in banner headlines that a Revolutionary War cannon ball had been discovered in the excavated garden behind our neighbor's yard. The ball was displayed for public view on the neighbor's mantel, to the mystification of visiting historians, who were unable to figure out the trajectory from any of the known engagements of the British or American forces; several learned papers were written on the problem. My father claimed privately to his family, swearing us to secrecy, that he had, in an indirect sense anyway, made medical history.

So far as I know, he was never caught up again by medical theory. He did not believe in focal infections when this notion appeared in the 1920s, and must have lost a lucrative practice by not removing normal tonsils, appendixes, and gallbladders. When the time for psychosomatic disease arrived, he remained a skeptic. He indulged my mother by endorsing her administration of cod-liver oil to the whole family, excepting himself, and even allowed her to give us something for our nerves called Eskay's Neurophosphates, which arrived as samples from one of the pharmaceutical houses. But he never convinced himself about the value of medicine.

His long disenchantment with medical therapy was gradually replaced by an interest in surgery, for which he found himself endowed with a special talent. At last, when he was in his early fifties, he decided to give up general practice and concentrate exclusively on surgery. He was very good at it, and his innate skepticism made him uniquely successful as a surgical consultant. Years later, after his death, I was told by some of his younger colleagues that his opinion was especially valued, and widely sought throughout the county, because of his known reluctance to operate on a patient until he was entirely convinced that the operation was absolutely necessary. His income must have suffered because of this, but his reputation was solidly established.

# 3
# 1911 MEDICINE

My father went to P&S in 1901, the College of Physicians and Surgeons of Columbia University, two years after he was graduated from Princeton. The education he received was already being influenced by the school of therapeutic nihilism for which Sir William Osler and his colleagues at Johns Hopkins had been chiefly responsible. This was in reaction to the kind of medicine taught and practiced in the early part of the nineteenth century, when anything that happened to pop into the doctor's mind was tried out for the treatment of illness. The medical literature of those years makes horrifying reading today: paper after learned paper recounts the benefits of bleeding, cupping, violent purging, the raising of blisters by vesicant ointments, the immersion of the body in either ice water or intolerably hot water, endless lists of botanical extracts cooked up and mixed together under the influence of nothing more than pure whim, and all these things were drilled into the heads of medical students—most of whom learned their trade as apprentices in the offices of older, established doctors. Osler and his colleagues introduced a revolution in medicine. They pointed out that most of the remedies in common use were more likely to do harm than good, that there were only a small number of genuine therapeutic drugs—digitalis and morphine the best of all, and they laid out a new, highly conservative curriculum for training medical students. By the time my father reached P & S, the principal concern of the faculty of medicine was the teaching of diagnosis. The recognition of specific illnesses, based on what had been learned about the natural history of disease and about the pathologic changes in each illness, was the real task of the doctor. If he could make an accurate diagnosis, he could forecast from this information what the likely outcome was to be for each of his patients' illnesses.

But medical students of those decades had other hard things to learn about. Prescriptions were an expected ritual, laid on as a kind of background music for the real work of the sixteen-hour day. First of all, the physician was expected to walk in and take over; he became responsible for the outcome whether he could affect it or not. Second, it was assumed that he would *stand by*, on call, until it was over. Third, and this was probably the most important of his duties, he would explain what had happened and what was likely to happen. All three duties required experience to be done well. The first two needed a mixture of intense curiosity about people in general and an inborn capacity for affection, hard to come by but indispensable for a good doctor. The third, the art of prediction, needed education, and was the sole contribution of the medical school; good medical schools produced doctors who could make an accurate diagnosis and knew enough of the details of the natural history of disease to be able to make a reliable prognosis. This was all there was to science in medicine, and the store of information which made diagnosis and prognosis possible for my father's generation was something quite new in the early part of the twentieth century.

The teaching hospitals of that time were organized in much the same way as now, although they existed on a much smaller scale than in today's huge medical centers. The medical school was responsible for appointing all the physicians and surgeons who worked on the wards, and these people held academic titles on the medical school faculty. The professor and head of the Department of Medicine in P & S was also the chief of the internal medicine service in Roosevelt Hospital, the professor of surgery ran the surgical service, the pediatrics professor was in charge of all the children's wards, and so forth. The medical students were assigned in rotation to each of the clinical services during the last two years of medical school. Interns were selected from applicants who were graduates of all of the country's medical schools, and the competition for appointments in the teaching hospitals was as intense then as it is today. To be posted as intern on one of the teaching services at Roosevelt Hospital was regarded as a sure ticket for a successful career as a practitioner in the New York City area. The P & S faculty included some of the city's leading physicians and surgeons, who made ward rounds each day with an entourage of students and interns, and taught their juniors everything they knew about medicine. Through this mechanism, the interns also had opportunities to observe, at first hand, some of the imperfections of medicine.

When my father was an intern, one of the attending physicians on the P & S medical service of Roosevelt Hospital was an elderly, highly successful pomposity of New York medicine, typical of the generation trained long before the influence of Sir William Osler. This physician enjoyed the reputation of a diagnostician, with a particular skill in diagnosing typhoid fever, then the commonest disease on the wards of New York's hospitals. He placed particular reliance on the appearance of the tongue, which was universal in the medicine of that day (now entirely inexplicable, long forgotten). He believed that he could detect significant differences by palpating that organ. The ward rounds conducted by this man were, essentially, tongue rounds; each patient would stick out his tongue while the eminence took it between thumb and forefinger, feeling its texture and irregularities, then moving from bed to bed, diagnosing typhoid in its earliest stages over and over again, and turning out a week or so later to have been right, to everyone's amazement. He was a more productive carrier, using only his hands, than Typhoid Mary.

Good doctors needed a close knowledge of what good nurses were able to do. In the medical textbooks used by my father, the sections dealing with the treatment of disease usually began with a paragraph about the importance of "good nursing care," and what this really meant was that the services of "good nurses" had to be available. The nurses had their own profession, their own schools, and their own secrets. It was understood that nurses were there to do what the physician told them to but he usually knew so little about nursing that his instructions were often no more precise than to "look after" the patient. Nurses knew how to make sick people more comfortable, and more confident and hopeful as well; they also knew how to get things done in hospitals.

My mother was trained in nursing at Roosevelt Hospital.

She had been raised on a small, always impoverished farm in Connecticut near Beacon Falls. She never talked much to her children about her own childhood, but we gathered enough to know that it had been hard times. She was orphaned at the age of five or six, raised by grandparents and several unaffectionate aunts; the farm was a spartan place; the aunts took her to church and then to the local graveyard every Sunday to memorize the names and dates of all the family antecedents, Pecks and Brewsters who, the aunts claimed, were Mayflower progeny (which my mother doubted, since everyone in that part of Connecticut was raised in the same belief); she fled when she could, at the age of seventeen, and boarded a boat in Bridgeport for New York. She carried a letter from the family doctor in New Haven to a colleague at Roosevelt Hospital, recommending her as a strong-minded, intelligent girl who would make a good nurse.

By the time my father had arrived at Roosevelt for his internship, she had finished nursing school and risen to head nurse on one of the wards. Later, she was elevated to the highest local distinction for a Roosevelt nurse, personal assistant to the chief surgeon, Dr. George Brewer. Brewer's practice took him to various estates on Long Island, and my mother's task was to go out a day in advance to prepare the household for his surgery, usually performed on a table in the kitchen. Wealthy people went through all their illnesses at home at the time; the hospital was seen as the place for dying.

My mother's skills were devoted almost exclusively to the family after she and my father were married. I can remember only a few occasions when my father served as the family doctor. When we developed serious illnesses he usually called one of his Flushing colleagues to make a house call to our house; it was considered vaguely unethical for a doctor to treat his own family. For most of our indispositions it was my mother's responsibility to look after us.

She took considerable pride in my father's accomplishments, probably more pride than he was comfortable with, but her view of him must have provided a lifelong reassurance whenever he ran into bad times. Around the house, talking with the family, she was entirely explicit: our father, we were told, was not simply the best doctor in town, not just the best in Queens County, he was very likely the best anywhere. I remember a time somewhere in the late 1920s when he had begun concentrating on surgery and had a particularly difficult case to deal with—the minister of the local Baptist church who had developed a severe gallbladder infection. My father operated on him one night at Flushing Hospital, and he recovered slowly after a few weeks. When he returned to his pulpit, fit again, he delivered a rousing sermon which was printed the next day on the front page of the *Evening Journal*, under the headline: BAPTIST MINISTER GIVES THANKS TO GOD FOR RECOVERY. My mother was indignant, flapping the newspaper impatiently on her lap. "God indeed!" she said to all of us. "God had nothing to do with it. It was your father!"

Once in a while, in emergencies, she pitched in to help my father in his office. The office door would open and we would hear him call, "Grace!" and we knew that something serious was happening downstairs, someone bleeding or fainting or needing more reassurance than he could provide. Some of his patients came back after office hours to see her, and as the years went by, she accumulated an informal, unpaid practice of her own.

One night my father was called out for an emergency and came home with a baby. He had arrived at the household of one of the old Flushing families a few minutes after the young daughter, unmarried, had delivered, and found the grandmother trying to smother the newborn child with a pillow. Mother looked after the baby for a few days, tried her best to negotiate with its family to accept the event, and finally brought it to her friends the nuns at the Foundling Hospital in New York. Later, I learned that this had happened several times in my father's practice. My mother, who had been raised in Protestant fundamentalism, told me once that the Catholic nuns at the Foundling Hospital were, to her surprise, the best nurses she'd ever seen.

A piano teacher, a middle-aged spinster, came to the office one morning hallucinating wildly and left in panic before my father came home for office hours. My mother took on the problem, visiting the lady's apartment on the other side of town twice each day with trays of hot food. From her account, it was a typical case of acute, severe schizophrenia, which in other circumstances would have required immediate removal to Bellevue or the asylum on Blackwells Island, but my mother insisted on handling the matter in her own way. For several weeks the piano teacher refused to open her door; she and my mother would discuss her problems for a while, and then the tray would be left on the threshold, to be taken in after Mother had left. This went on for a couple of months, and then the lady recovered and resumed her successful career. All of this went on in secrecy; nobody in town knew anything more than that the music teacher was "ill" at home and my mother was "looking in" at her apartment.

# 4
# 1933 Medicine

I was admitted to medical school under circumstances that would have been impossible today. There was not a lot of competition; not more than thirty of my four hundred classmates, most of these the sons of doctors, planned on medicine. There was no special curriculum; elementary physics and two courses in chemistry were the only fixed requirements; the term "premedical" had not yet been invented. My academic record at Princeton was middling fair; I had entered college at fifteen, having been a bright enough high-school student, but then I turned into a moult of dullness and laziness, average or below average in the courses requiring real work. It was not until my senior year, when I ventured a course in advanced biology under Professor Swingle, who had just discovered a hormone of the adrenal cortex, that I became a reasonably alert scholar, but by that time my grade averages had me solidly fixed in the dead center, the "gentlemen's third," of the class. Today, I would have been turned down by every place, except perhaps one of the proprietary medical schools in the Caribbean.

I got into Harvard, the hardest, by luck and also, I suspect, by pull. Hans Zinsser, the professor of bacteriology, had interned with my father at Roosevelt and had admired my mother, and when I went to Boston to be interviewed in the winter of 1933, I was instructed by the dean's secretary to go have a talk with Dr. Zinsser. It was the briefest of interviews, but satisfactory from my point of view. Zinsser looked at me carefully, as at a specimen, then informed me that my father and mother were good friends of his, and if I wanted to come to Harvard he would try to help, but because of them, not me; he was entirely good-natured, but clear on this point. It was favoritism, but not all that personal, I was to understand.

My medical education was, in principle, much like that of my father. The details had changed a lot since his time, especially in the fields of medical science relating to disease mechanisms; physiology and biochemistry had become far more complex and also more illuminating; microbiology and immunology had already, by the early 1930s, transformed our understanding of the causation of the major infectious diseases. But the *purpose* of the curriculum was, if anything, even more conservative than thirty years earlier. It was to teach the recognition of disease entities, their classification, their signs, symptoms, and laboratory manifestations, and how to make an accurate diagnosis. The treatment of disease was the most minor part of the curriculum, almost left out altogether. There was, to be sure, a course in pharmacology in the second year, mostly concerned with the mode of action of a handful of everyday drugs: aspirin, morphine, various cathartics, bromides, barbiturates, digitalis, a few others.

Vitamin B was coming into fashion as a treatment for delirium tremens, later given up. We were provided with a thin, pocket-size book called *Useful Drugs*, one hundred pages or so, and we carried this around in our white coats when we entered the teaching wards and clinics in the third year, but I cannot recall any of our instructors ever referring to this volume. Nor do I remember much talk about treating disease at any time in the four years of medical school except by the surgeons, and most of their discussions dealt with the management of injuries, the drainage or removal of infected organs and tissues, and, to a very limited extent, the excision of cancers.

The medicine we were trained to practice was, essentially, Osler's medicine. Our task for the future was to be diagnosis and explanation. Explanation was the real business of medicine. What the ill patient and his family wanted most was to know the name of the illness, and then, if possible, what had caused it, and finally, most important of all, how it was likely to turn out.

The successes possible in diagnosis and prognosis were regarded as the triumph of medical science, and so they were. It had taken long decades of careful, painstaking observation of many patients; the publication of countless papers describing the detailed aspects of one clinical syndrome after another; more science, in the correlation of the clinical features of disease with the gross and microscopic abnormalities, contributed by several generations of pathologists. By the 1930s we thought we knew as much as could ever be known about the dominant clinical problems of the time: syphilis, tuberculosis, lobar pneumonia, typhoid, rheumatic fever, erysipelas, poliomyelitis. Most of the known varieties of cancer had been meticulously classified, and estimates of the duration of life could be made with some accuracy. The electrocardiogram had arrived, adding to the fair precision already possible in the diagnosis of heart disease. Neurology possessed methods for the localization of disease processes anywhere in the nervous system. When we had learned all that, we were ready for our M.D. degrees, and it was expected that we would find out about the actual day-to-day management of illness during our internship and residency years.

During the third and fourth years of school we also began to learn something that worried us all, although it was not much talked about. On the wards of the great Boston teaching hospitals—the Peter Bent Brigham, the Massachusetts General, the Boston City Hospital, and Beth Israel—it gradually dawned on us that we didn't know much that was really useful, that we could do nothing to change the course of the great majority of the diseases we were so busy analyzing, that medicine, for all its façade as a learned profession, was in real life a profoundly ignorant occupation.

Some of this we were actually taught by our clinical professors, much more we learned from each other in late-night discussions. When I am asked, as happens occasionally, which member of the Harvard faculty had the greatest influence on my education in medicine, I no longer grope for a name on that distinguished roster. What I remember now, from this distance, is the influence of my classmates. We taught each other; we may even have set careers for each other without realizing at the time that so fundamental an educational process was even going on. I am not so troubled as I used to be by the need to reform the medical school curriculum. What worries me these days is that the curriculum, whatever its sequential arrangement, has become so crowded with lectures and seminars, with such masses of data to be learned, that the students may not be having enough time to instruct each other in what may lie ahead.

The most important period for discovering what medicine would be like was a three-month ward clerkship in internal medicine that was a required part of the fourth year of medical school. I applied for the clerkship at the Beth Israel Hospital, partly because of the reputation of Professor Hermann Blumgart and partly because several of my best friends were also going there. Ward rounds with Dr. Blumgart were an intellectual pleasure, also good for the soul. I became considerably less anxious about the scale of medical ignorance as we followed him from bed to bed around the open circular wards of the B.I. I've seen his match only three or four times since then. He was a tall, thin, quick-moving man, with a look of high intelligence, austerity, and warmth all at the same time. He had the special gift of perceiving, almost instantaneously, while still approaching the bedside of a new patient, whether the problem was a serious one or not. He seemed to do this by something like intuition; at times when there were no particular reasons for alarm that could be sensed by others in the retinue, Blumgart would become extremely alert and attentive, requiring the resident to present every last detail of the history, and then moving closer to the bedside, asking his own questions of the patient, finally performing his physical examination. To watch a master of physical diagnosis in the execution of a complete physical examination is something of an aesthetic experience, rather like observing a great ballet dancer or a concert cellist. Blumgart did all this swiftly, then asked a few more questions, then drew us away to the corridor outside the ward for his discussion, and then his diagnosis, sometimes a death sentence. Then back to the bedside for a brief private talk with the patient, inaudible to the rest of us, obviously reassuring to the patient, and on to the next bed. So far as I know, from that three months of close contact with Blumgart for three hours every morning, he was never wrong, not once. But I can recall only three or four patients for whom the diagnosis resulted in the possibility of doing something to change the course of the illness, and each of these involved calling in the surgeons to do the something—removal of a thyroid nodule, a gallbladder, an adrenal tumor. For the majority, the disease had to be left to run its own course, for better or worse.

There were other masters of medicine, each as unique in his way as Blumgart, surrounded every day by interns and medical students on the wards of the other Boston hospitals.

The Boston City Hospital, the city's largest, committed to the care of indigent Bostonians, was divided into five separate clinical services, two staffed by Harvard Medical School (officially designated as the Second and Fourth services), two by Tufts, and one by Boston University. The most spectacular chiefs on the Harvard faculty were aggregated on the City Hospital wards, drawn there in the 1920s by the creation of the Thorndike Memorial Laboratories, a separate research institute on the hospital grounds, directly attached by a series of ramps and tunnels to the buildings containing the teaching wards. The Thorndike was founded by Dr. Francis Weld Peabody, still remembered in Boston as perhaps the best of Harvard physicians. Peabody was convinced that the study of human disease should not be conducted solely by bedside observations, as had been largely the case for the research done by physicians up to that time, nor by pure bench research in the university laboratories; he believed that the installation of a fully equipped research institute, containing laboratories for investigations of any promising line of inquiry, directly in communication with the hospital wards, offered the best opportunity for moving the field forward.

Peabody was also responsible for the initial staffing of the Thorndike. By the time I arrived, in 1937, the array of talent was formidable: George Minot (who had already received his Nobel prize for the discovery of liver extract as a cure for pernicious anemia), William Castle (who discovered the underlying deficiency in pernicious anemia), Chester Keefer, Soma Weiss, Maxwell Finland, John Dingle, Eugene Stead—each of them running a laboratory, teaching on the wards, and providing research training for young doctors who came for two- or three-year fellowship stints from teaching hospitals across the country. The Thorndike was a marvelous experiment, a model for what were to become the major departments of medicine in other medical schools, matched at the time only by the hospital of the Rockefeller Institute in New York.

Max Finland built and then ran the infectious disease service. He and his associates had done most of the definitive work on antipneumococcal sera in the treatment of lobar pneumonia, testing each new preparation of rabbit antiserum as it arrived from the Lederle Laboratories. Later, Finland's laboratories were to become a national center for the clinical evaluation of penicillin, streptomycin, chloromycetin, and all the other antibiotics which followed during the 1950s and 1960s. As early as 1937, medicine was changing into a technology based on genuine science. The signs of change were there, hard to see because of the overwhelming numbers of patients for whom we could do nothing but stand by, but unmistakably there all the same. Syphilis could be treated in its early stages, and eventually cured, by Paul Ehrlich's arsphenamine; the treatment took a long time, many months, sometimes several years. If arsphenamine was started in the late stages of the disease, when the greatest damage was under way—in the central nervous system and the major arteries—the results were rarely satisfactory—but in the earliest stages, the chancre and then the rash of secondary syphilis, the spirochete could be killed off and the Wassermann reaction reversed. The treatment was difficult and hazardous, the side effects of the arsenical drugs were appalling, sometimes fatal (I cannot imagine such a therapy being introduced and accepted by any of today's FDA or other regulatory agencies), but it did work in many cases, and it carried a powerful message for the future: it was possible to destroy an invading microorganism, intimately embedded within the cells and tissues, without destroying the cells themselves. Chemotherapy for infectious disease in general lay somewhere ahead, and we should have known this.

Immunology was beginning to become an applied science. Thanks to the basic research launched twenty years earlier by Avery, Heidelberger, and Goebbel, it was known that pneumococci possessed specific carbohydrates in their capsules which gave rise to highly specific antibodies. By the mid-1930s, rabbit antipneumococcal sera were available for the treatment of the commonest forms of lobar pneumonia. The sera were difficult and expensive to prepare, and sometimes caused overwhelming anaphylactic reactions in patients already moribund from their infection, but they produced outright cures in many patients. Pernicious anemia, a uniformly fatal disease, was spectacularly reversed by liver extract (much later found to be due to the presence of vitamin $B_{12}$ in the extracts). Diabetes mellitus could be treated—at least to the extent of reducing the elevated blood sugar and correcting the acidosis that otherwise led to diabetic coma and death—

by the insulin preparation isolated by Banting and Best. Pellagra, a common cause of death among the impoverished rural populations in the South, had become curable with Goldberger's discovery of the vitamin B complex and the subsequent identification of nicotinic acid. Diphtheria could be prevented by immunization against the toxin of diphtheria bacilli and, when it occurred, treated more or less effectively with diphtheria antitoxin.

All these things were known at the time of my internship at the Boston City Hospital, but they seemed small advances indeed. The major diseases, which filled the wards to overflowing during the long winter months, were infections for which there was no treatment at all.

The two great hazards to life were tuberculosis and tertiary syphilis. These were feared by everyone, in the same way that cancer is feared today. There was nothing to be done for tuberculosis except to wait it out, hoping that the body's own defense mechanisms would eventually hold the tubercle bacillus in check. Some patients were helped by collapsing the affected lung (by injecting air into the pleural space, or by removing the ribs overlying the lung), and any number of fads were introduced for therapy—mountain resorts, fresh air, sunshine, nutritious diets—but for most patients tuberculosis simply ran its own long debilitating course despite all efforts. Tertiary syphilis was even worse. The wards of insane asylums were filled with psychotic patients permanently incapacitated by this disease—"general paresis of the insane"; some benefit was claimed for fever therapy; but there were few real cures. Rheumatic fever, the commonest cause of fatal heart disease in children, was shown by Coburn to be the result of infection by hemolytic streptococci; aspirin, the only treatment available, relieved the painful arthritis in this disease but had no effect on the heart lesions. For most of the infectious diseases on the wards of the Boston City Hospital in 1937, there was nothing to be done beyond bed rest and good nursing care.

Then came the explosive news of sulfanilamide, and the start of the real revolution in medicine.

I remember the astonishment when the first cases of pneumococcal and streptococcal septicemia were treated in Boston in 1937. The phenomenon was almost beyond belief. Here were moribund patients, who would surely have died without treatment, improving in their appearance within a matter of hours of being given the medicine and feeling entirely well within the next day or so.

The professionals most deeply affected by these extraordinary events were, I think, the interns. The older physicians were equally surprised, but took the news in stride. For an intern, it was the opening of a whole new world. We had been raised to be ready for one kind of profession, and we sensed that the profession itself had changed at the moment of our entry. We knew that other molecular variations of sulfanilamide were on their way from industry, and we heard about the possibility of penicillin and other antibiotics; we became convinced, overnight, that nothing lay beyond reach for the future. Medicine was off and running.

# 5
# 1937 INTERNSHIP

No job I've ever held since graduating from medical school was as rewarding as my internship. Rewarding may be the wrong word for it, for the salary was no money at all. A bedroom, board, and the laundering of one's white uniform were provided by the hospital; the hours of work were all day every day, and on call for admissions and emergencies every other night, all night long. There was no such thing as a weekend. The hours were real working hours; when the night came, especially in the winter months, the intern was even more on the run than during the daytime shift.

I am remembering the internship through a haze of time, cluttered by all sorts of memories of other jobs, but I haven't got it wrong nor am I romanticizing the experience. It was, simply, the best of times.

The lack of money was no great problem. None of the interns on the two Harvard Medical services at the Boston

City Hospital was married. It would have been unheard of, and very likely a married applicant for this internship would have been rejected by the selection committee just for being married. There was little need for pocket money because there was no time to spend pocket money. In any case the interns had one sure source of spare cash: they were the principal donors of blood transfusions, at $25 a pint; two or three donations a month kept us in affluence. More than this, Massachusetts law in 1937 stipulated that a blood donor was entitled to a pint of whiskey; at the Boston City Hospital the pint was Golden Wedding.

The internship was divided into six periods of three months each. One rose through the hierarchy automatically, but the jumps from one rank to the next seemed quantum leaps. The newest man was the Junior, also known as the pup; his life was spent collecting specimens of blood, urine, feces, spinal fluid, sputum, sometimes pleural fluid, and doing the laboratory diagnostic work—all the work for his assigned wards of thirty patients each. The day never ended; it was one long twenty-four-hour run, trying to keep up with the orders coming down from the upper ranks. The second three months was the externship, with two major responsibilities: all morning in the outpatient department, mostly sitting at a desk listening to the complaints of elderly patients who liked to spend their mornings visiting the clinics of the City Hospital; very few of them were acutely ill, many of them were patients who had been discharged from the hospital a few weeks earlier and were back for checkups; some of the patients had chronic ailments—diabetes, arthritis, hypertension, mild heart failure. The other half of the extern's day was spent back on the wards performing the therapeutic procedures, such as they were, the installations of intravenous drips of saline, transfusions, injections of insulin or liver extract, and the administration of antipneumococcal antiserum. The third three-month period was spent across the street in a huge annex called the South Department, where all the contagious diseases were cared for—several hundred patients, mostly children, with diphtheria, whooping cough, scarlet fever, chicken pox, measles, and poliomyelitis. The work here was the same as on the general wards, but more time-consuming because of the necessity of wearing sterile gowns, gloves, and masks, and changing these from one bedside to the next.

The last nine months contained the reward for the first nine: the privilege of giving the orders instead of taking them. The Senior had the responsibility for admitting patients to the ward, taking their histories, doing the physical examinations, deciding what laboratory tests and therapeutic procedures were to be ordered.

The next rank up, the Assistant House Physician, supervised the Senior and went off on consultation calls to the surgical, neurological, and psychiatric services during most of his days and nights on duty. The top of the hierarchy was the House Physician, equivalent in rank and prestige to a Chief Resident on today's house staff—except that one became House automatically after fifteen months of duty, while the contemporary Chief Resident is selected from among competitors at the end of four or five years.

All day long the visiting physicians, in long white coats, moved back and forth across the ramp connecting the wards of the Peabody Building (which housed the Fourth Division, one of the two Harvard Medical services) and their laboratories and offices in the Thorndike. They came at ten in the morning to make the formal rounds, walked the bedsides for two or three hours with the interns and medical students, and then came back at odd hours throughout the afternoon and often until late in the evening to see patients with serious problems in whom they were especially interested.

The wards were long, high-ceilinged rectangles, with thirty beds around the periphery. In winter each ward filled and doubled its capacity by adding thirty cots, lined up in two rows down the middle of the room. Ward rounds at this season took a longer time than in summer, partly because of the greater number of very sick patients, but also because of the logistic problem of moving the entourage of physicians and nurses over and around the beds and bed tables jumbled into every available space.

Each ward possessed a single private room off at one end, reserved, in theory anyway, for the dying. In practice, it was used for patients in delirium whose outcries in the night would have kept the other patients awake. Dying was done out on the wards, and it happened every day and every night. Each bed was surrounded by poles carrying white curtains on metal rings; these were pulled to surround the bed for the privacy of physical examinations and for the time of dying. When a patient was dead, the removal of the body from the ward was accomplished by a ceremony with which all the new patients quickly became familiar. The head nurse would start at one end and walk rapidly from bed to bed, pulling the front curtain across the foot of each. Wherever you were in that building you could hear the zing-zing of the curtain rings, twenty-nine of them, and then the stretcher with the body would be rolled out of its cubicle, along the ward, out to the elevator, and down to the morgue. The patients on the cots in the center of the ward were hidden by movable screens.

Everyone on the ward knew what was going on. It was not an effort to protect the sensibilities of the living, it was done in respect for the privacy of the newly dead.

The open ward was viewed as a necessity, so that all patients could be within sight of the ward nurse at all hours. The public wards of hospitals of the time were run on very low budgets, and it was only on the private services of the great voluntary hospitals of Boston that a patient could afford a private room and a private nurse. Yet few people on the open wards complained; they made friends with each other quickly, and those who were well enough to be up and around moved from neighboring bed to bed, companionably gossiping, helping in the feeding of patients too sick to manage for themselves.

The greatest part of the work done on these wards was simply custodial. Patients came in with their illnesses, almost all of them severe; you did not go to the emergency room for admission to the Boston City Hospital unless you felt in danger for your life. Once you were admitted, transported on wheeled litters through the tunnels of the hospital, up the elevators, and onto the wards, it became a matter of waiting for the illness to finish itself one way or the other. If being in a hospital bed made a difference, it was mostly the difference produced by warmth, shelter, and food, and attentive, friendly care, and the matchless skill of the nurses in providing these things. Whether you survived or not depended on the natural history of the disease itself. Medicine made little or no difference.

And yet, everyone, all the professionals, were frantically busy, trying to cope, doing one thing after another, all day and all night. Most of this effort was aimed at being certain that nothing was missed, that the diagnosis was a matter of certainty, and that the illness was not one of the few for which there was believed to be a genuine and effective treatment.

There were only a few, and they became the great emergencies of the ward when they were recognized.

The commonest one, and the illness requiring the hardest and most urgent work by the intern, was lobar pneumonia. The pneumonia season began in late autumn and lasted until early spring. In order to make sure that no ward and no single intern was overburdened by the work, there was a system called the "pneumonia count"; the cases were sent up from the emergency room in rotation among the Harvard, Tufts, and BU medical services; if an intern was suddenly confronted by four patients with lobar pneumonia late at night, he was in for it, but at least he knew that each of his colleagues on the other services had the same problem. The diagnosis was usually the simplest part; the patient complained of the sudden onset of chills and fever, cough, sometimes with blood-tinged sputum, and pain in one side of the chest; physical examination revealed dullness to percussion with one's fingertips over the affected lung area and a characteristic change in the breath sounds heard with the stethoscope at the same spot. Given this amount of information you could begin making predictions. The prognosis for a young adult was the most surely predictable: an acute illness lasting ten to fourteen days, with a high fever each day, more chest pain and more cough, perhaps with alarming manifestations of exhaustion and debilitation near the end of this period, and then, suddenly and as triumphantly as the bright sunshine after a thunderstorm, one of the great phenomena of human disease—the *crisis*. On one day or another, after two weeks of his seeming to come closer and closer to death's door, the patient's temperature would drop precipitously within a few hours from 106 degrees to normal, and at the same time, with a good deal of sweating, the patient would announce that he felt better now and would like something to eat, and the illness would end, like that. All of this could be nicely explained; indeed, lobar pneumonia is the only disease I can remember that had an intellectually satisfying explanation at that time for what went on. The cause was the pneumococcus, a bacterium which was stained dark blue by the gram stain, always as two round, paired cocci. The capsule of this organism contained a polysaccharide, a carbohydrate that endowed it with its invasive properties and protected it against being engulfed and killed off by the host's white blood cells. There were about forty different types of pneumococcus, each with its own special type of capsular polysaccharide. The game to be played involved, initially, the invasion of the alveolar spaces of the lung by the pneumococcus, the spreading of new generations of pneumococci through that part of the lung, until an entire lobe became solidified by the jammed populations of bacteria and still-ineffectual leukocytes within the alveoli, sometimes the spread of the organism into the patient's bloodstream with septicemia, and then, on or around the tenth day, the mobilization of an effective defense by the patient's own antibody, chemically designed to fit precisely with the molecular configuration of the polysaccharides of that strain of pneumococcus and no other. Once this happened, and the levels of circulating antibody in the blood were high enough to have combined with all of the polysaccharides, the pneumococcus was the loser. When it combined with the antibody it was immediately swept up by the leukocytes and killed, and the disease was over. This event was the crisis, the sudden drop of the temperature, the sweats, the return of appetite, the end of the game.

The odds were shifted by several circumstances which we knew about and which changed the sense of urgency. Some types of pneumococcus were more virulent than others, needing quicker treatment. Patients known to be alcoholics were much more vulnerable to septicemia and overwhelming infection than normal people. Pregnant women were more susceptible and at greater risk of dying. Old people were the greatest risks of all.

The treatment was designed to match what we knew of the disease mechanism: the intravenous administration of type-specific antibody directed against the polysaccharide of the particular pneumococcus. Commercial preparations of purified rabbit antibodies against most of the known strains of pneumococci were available in the Thorndike laboratories, and it was the intern's first and most urgent task to identify the pneumococcus so that the proper serum could be used. This could be done with a specimen of sputum, in which the pneumococci were usually present in abundance. One simply added samples of various antipneumococcal sera to bits of sputum and stained them with methylene blue; if you had the right serum, the capsule around each of the paired organisms would become swollen and dark blue. If you were lucky, and had a good sample of sputum (saliva wouldn't do, it had to be real, coughed-up sputum), and had the right diagnostic serum at hand, you could make the diagnosis within a few minutes and telephone Dr. Finland's laboratory for the needed supply of therapeutic serum. If not lucky, you had to wait; the blood culture might grow out of the pneumococcus within the next two days and you could type it then, or you could inject the sputum sample into a white mouse, and if there were pneumococci there that you'd missed seeing, they would grow out in the peritoneal cavity of the mouse within a few hours, and you could learn the type that way. One way or another, the intern *had* to find out the type; there could be no going to bed until that job had been successfully done. The treatment depended on the precise answer, and no other answer would do; it was not enough to know that the diagnosis was lobar pneumonia, not enough to know that the organisms were pneumococci; you needed to know next whether it was a type I, or a type III, or a type whatever, or there would be no way at all of treating the disease, nothing at all to do beyond watching the illness run its natural course.

But when the type was known, the treatment could turn out a technological *tour de force*, the only genuine cure within an intern's grasp.

Given knowledge of the type, and a supply of the right rabbit serum, the intern became a man of power. The serum was injected, very slowly, by vein. When it worked, it worked within an hour or two. Down came the temperature, and the patient, who might have been moribund a few hours earlier, would now be sleeping in good health.

It didn't always come out this way, but it was successful often enough to make it worth great effort. An intern was judged by his superiors on this kind of success more than by any other quality: if your lobar pneumonia cases were well handled, you were likely to have a future; if not, not.

The second of the great emergencies, even more demanding of quick thinking and precise action, was diabetic coma. If it was recognized early enough and the right things done with speed, an intern could have the certainty of saving a life, but if he stalled around or miscalculated the need for insulin and intravenous fluids, the patient might die. The management of coma was one of the occasions for calling on all the help and advice available. The senior visiting physicians came across the ramp from the Thorndike on the run, medical students were summoned from their quarters, and the house staff, all levels of interns, gathered at the bedside to work out the details together.

Acute heart failure was another. There were only three things to be done, three sorts of technology, not always effective and never in any sense curative, but when done properly they possessed near-magical properties for pulling an occasional patient back from dying. Bleeding was the first, the rapid withdrawal of a pint of blood from an arm vein; often enough this served by itself to take away the desperate gasping for breath by reducing the load of venous blood to be coped with by the heart. Digitalis was the second, given gradually and with gingerly caution, just enough of the crude leaf preparations then available to add strength to the heart muscle, not enough to cause toxicity—it was a skill long mastered by the older clinicians on the visiting staff, very difficult to learn from scratch, something like cooking. The third was oxygen, provided by a tank brought to the bedside, delivered by way of an oxygen tent if one was available, otherwise by a rubber tube fixed in the nose by adhesive tape.

And then there was syphilis. This was never an emergency for therapy; whenever it reached one or another of its life-endangering phases—an aortic aneurysm about to burst, for example, or a brain disintegrating from paresis—it was already too late for anything really effective. The time for action, and for diagnostic acumen in recognizing the disease, was in its earliest stages. An intern's heart sank at the thought of what then lay ahead: months, even years, of arsenicals, mercury, and bismuth, with risks of liver destruction from the treatment itself. In the old days, at the Boston City Hospital, you had to rule out this disease, called "the great imitator," before proceeding on other diagnostic lines. It was the routine to look first for the Argyll-Robertson pupil; you did this by two swift maneuvers: first you aimed your pocket flashlight at the eye and observed it for constriction of the pupil, then you persuaded the patient to look at your finger moving closer to the eye and watched for the pupil to constrict in accommodation; if it failed the first test but responded to the second you had your diagnosis: neurosyphilis. I don't suppose a busy internist sees an Argyll-Robertson pupil more than once a year these days. The disease has vanished, for the time anyway. There is still syphilis, of course, probably more than at any time in the history of society due to the new sexual freedom, but it is easily cut off in its primary or secondary stages, and very few patients move on to the remorseless and lethal third stage, which catches the brain. No credit, or not much credit anyway, to medicine as a profession. Indeed, this entirely commendable advance in public health is very likely the result of bad medicine, the indiscriminate use of penicillin for any minor ailment, a cough, a head cold. The spirochete of syphilis has been caught up in a national aerosol of penicillin, and can only rarely gasp its way through to the formation of a chancre. Paresis and tabes, which used to be the principal anxiety of any young intern, have departed. No longer is a clinical clerk in Boston required to search for Argyll-Robertson pupils or to ask the patient to stand with his feet together and shut his eyes (the swaying and falling indicated tabes), or, as I was taught, to command the patient to repeat "God save the

Commonwealth of Massachusetts." In 1937, if you were obviously deranged but could manage that sentence, you had schizophrenia; if you stumbled over it you had paresis.

Boston had lots of alcoholics in the 1930s, and the City Hospital contained a special ward of around forty beds just for them. Any night there would be three or four patients sent up from the emergency room with delirium tremens, as bad a sight to look at as anything in an intern's life experience: tremulous, hallucinating, wild-eyed, unhinged men. The risk was fever; some of these patients would suddenly run their body temperature up to 108 or over, and then die in an aftermath of deep shock. The treatment was paraldehyde, huge doses by mouth, sometimes even injected into the buttock muscles, enough to produce near-anaesthesia. Ice packs at hand in case of the fever, vitamin B, and liver extract, also in huge doses (although nobody was at all sure what good these did, if any). That was it for the DTs, and we saw a lot of deaths.

The Peabody Building had its working center in the laboratory on the top floor. Here were the microscopes and other equipment for doing blood counts and sedimentation rates, the urines and stools lined up for the Junior's inspection each morning, the incubators for culturing samples of blood, pleural fluid, and spinal fluid. For the more specialized chemical tests—blood sugar, cholesterol, several tests for abnormal serum proteins believed useful in diagnosing cirrhosis, the blood levels of nonprotein nitrogen (up in kidney failure)—the blood specimens were carried by the intern to the hospital's central diagnostic laboratory, which occupied a couple of small rooms in the basement of an adjacent building. The complicated bacteriological procedures, especially those needed for diagnosing meningitis, pneumonia, and septicemia, were performed by one or another of the Thorndike research laboratories.

The Peabody laboratory was the meeting place for the interns and medical students. A huge record book lay on a table just inside the door, containing the name and preliminary diagnosis of every patient admitted to the wards below. Traffic in and out of the door was rapid and incessant, people bumping into each other, carrying racks of test tubes or arms filled with reference journals.

The Peabody had one secretary in a tiny office just off the laboratory, who typed each day the detailed case summaries for all the patients who were being discharged, and especially long and meticulous summaries for the patients who had died. Each Friday afternoon the house staff and visiting physicians met with one of the hospital pathologists, and the day-to-day records of all patients who had died were combed through, looking for mistakes.

The worst mistake of all—perhaps the worst mistake I've ever observed at first hand—occurred in the first month of my service as Junior. A young black musician was admitted with a history of severe chills and fever during the preceding week. It was pneumonia season and the Senior intern suspected lobar pneumonia but could find nothing on physical examination. I was responsible for trying to get a sputum sample from him at the time of taking blood for the routine blood counts. The patient was drowsy and apathetic, had no cough, could raise no sputum, and I went along with my tray to collect the other blood specimens listed for the ward, then later in the morning upstairs to the laboratory to do the tests. His hemoglobin was alarmingly low, and I telephoned the ward with the news that the new patient had acute anemia; then I looked at the blood smear. I'd never seen anything like it. Almost every red cell contained blue-staining bodies, looking exactly like the textbook pictures of malaria. The hematologists came running across the ramp to take a look, then down to the ward to collect their own samples of blood. Pretty soon everyone arrived, all the house staff, all the visiting physicians, all the students. Nobody in Boston had ever seen malaria, it seemed.

Later in the day it made sense. The patient admitted to being a heroin addict, accustomed to parties where the needle and syringe were passed around; someone in the group, probably someone from far out of town, must have been the source.

Meanwhile, the ward visits continued all afternoon. The patient became drowsier, then, early in the evening, deeply comatose, and died within the next hour. He had the most malignant form of malaria, with clumps of infected cells occluding the small blood vessels of his brain. Had he received less clinical interest and animated attention, and been given quinine immediately the diagnosis was made, early that morning, he would perhaps have lived. The opportunity to cure an illness, even save a life, came infrequently enough on the City Hospital wards. This one had come and gone. The house physician went to his room and brought back his copy of Osler's *Textbook of Medicine*, opened to the chapter on malaria. The first sentence, which he read aloud to the assembled house staff, said, in effect: Any doctor who allows a case of malaria to die without quinine is guilty of malpractice.

The Fourth Medical Division regarded itself as the top service at the Boston City Hospital, the elite in the sharpest and brightest of all the teaching hospitals in town. We were the iron men, we told ourselves. The lights in the laboratory on the top floor of the Peabody Building were never turned off at night; the house staff never slept. After a while, we managed to talk ourselves out of the deep guilt which we had earned and into the easier sense of having been humiliated. Cerebral malaria, we agreed with ourselves, often moves so fast that nothing, no dose of quinine, could have made a difference. But the memory lasted: those clusters of white-uniformed professionals moving back and forth from the bedside of that extremely interesting case, taking new blood samples, discussing and discussing, and doing, in the end, nothing. It was a bad day for Harvard.

# 6
# *Leech Leech, Et Cetera*

A few years ago, I blundered into the fringes of a marvelous field of scholarship, comparative philology. I wondered—I forget the occasion—why leech was the word for the doctor and at the same time for the worm used by the doctor for so many centuries. Which came first, leech the doctor or leech the worm?

The lovely *American Heritage Dictionary* has a fifty-page appendix of Indo-European roots, based in large part on *Pokorny's Dictionary of Indo-European Languages*. My wife searched New York's bookstores and found a copy of *Pokorny* in a rare-book store for my birthday, and I have never since looked back.

The evolution of language can be compared to the biological evolution of species, depending on how far you are willing to stretch analogies. The first and deepest question is open and unanswerable in both cases: how did life start up at its very beginning? What was the very first human speech like?

Fossils exist for both, making it possible to track back to somewhere near the beginning. The earliest forms of life were the prokaryotes, organisms of the same shape and size as bacteria; chains of cocci and bacilli left unmistakable imprints within rocks dating back as far as 3.5 billion years. Similar microorganisms comprised the total life of the planet for the next 2.5 billion years, living free or, more often, gathered together as immense colonies in "algal mats," which later on fossilized into the formidable geological structures known as stromatolites. It was only recently, perhaps a billion years ago, that the prokaryotic algae had pumped enough oxygen into the earth's atmosphere so that nucleated cells could be formed. The mitochondria, which provide oxidative energy for all nucleated cells, and the chloroplasts of plant cells, which engage the sun's energy for producing the planet's food and oxygen, are the lineal descendants of bacteria and blue-green algae, and have lived as symbionts with the rest of us for a billion years.

The fossils of human language are much more recent, of course, and can only be scrutinized by the indirect methods of comparative philology, but they are certainly there. The most familiar ones are the Indo-European roots, prokaryote equivalents, the ancestors of most of the Western and some of the Eastern languages: Sanskrit, Greek, Latin, all the Slavic and Germanic tongues, Hittite, Tocharian, Iranian, Indic, some others, all originating in a common speech more than 20,000 years ago at a very rough guess. The original words from which the languages evolved were probably, at the outset, expressions of simple, non-nucleated ideas, unambiguous etymons.

The two leeches are an example of biological mimicry at

work in language. The root for leech the doctor goes back to the start of language: *leg* was a word meaning "to collect, with derivatives meaning to speak" and carried somehow the implication of knowledge and wisdom. It became *laece* in Old English, *lake* in Middle Dutch, with the meaning of doctor. Along the way, in early Germanic, it yielded *lekjaz*, a word meaning "an enchanter, speaking magic words," which would fit well with the duties of early physicians. The doctor was called the leech in English for many centuries, and a Danish doctor is still known as *Laege*, a Swedish one as *Lakere*.

*Leg* gave spawn to other progeny, different from the doctor but with related meanings. Lecture, logic, and logos are examples to flatter medicine's heart.

Leech the worm is harder to trace. The *OED* has it in tenth-century records as *lyce*, later *laece*, and then the two leeches became, for all practical purposes, the same general idea. Leech the doctor made his living by the use of leech the worm; leech the worm was believed (wrongly, I think) to have had restorative, health-giving gifts and was therefore, in its way, a sort of doctor. The technical term "assimilation" is used for this fusion of words with two different meanings into a single word carrying both. The idea of collecting has perhaps sustained the fusion, persisting inside each usage: blood for the leech, fees (and blood as well) for the doctor. Tax collectors were once called leeches, for the worm meaning, of course.

The word doctor came from *dek*, meaning something proper and acceptable, useful. It became *docere* in Latin, to teach, also *discere*, to learn, hence disciple. In Greek it was understood to mean an acceptable kind of teaching, thus dogma and orthodox. Decorum and decency are cognate words.

Medicine itself emerged from root *med*, which meant something like measuring out, or taking appropriate measures. Latin used *med* to make *mederi*, to look after, to heal. The English words moderate and modest are also descendants of *med*, carrying instructions for medicine long since forgotten; medical students ought to meditate (another cognate) from time to time about these etymological cousins.

The physician came from a wonderful word, one of the master roots in the old language, *bheu*, meaning nature itself, being, existence. *Phusis* was made from this root in Greek, on its way to the English word physic, used for medicine in general, and physics, meaning the study of nature.

Doctor, medicine, and physician, taken together with the cognate words that grew up around them, tell us a great deal about society's ancient expectations from the profession, hard to live up to. Of all the list, moderate and modest seem to me the ones most in need of remembering. The root *med* has tucked itself inside these words, living as a successful symbiont, and its similar existence all these years inside medicine should be a steady message for the teacher, the healer, the collector of science, the old leech.

Medicine was once the most respected of all the professions. Today, when it possesses an array of technologies for treating (or curing) diseases which were simply beyond comprehension a few years ago, medicine is under attack for all sorts of reasons. Doctors, the critics say, are applied scientists, concerned only with the disease at hand but never with the patient as an individual, whole person. They do not really listen. They are unwilling or incapable of explaining things to sick people or their families. They make mistakes in their risky technologies; hence the rapidly escalating cost of malpractice insurance. They are accessible only in their offices in huge, alarming clinics or within the walls of terrifying hospitals. The word "dehumanizing" is used as an epithet for the way they are trained, and for the way they practice. The old art of medicine has been lost, forgotten.

The American medical schools are under pressure from all sides to bring back the family doctor—the sagacious, avuncular physician who used to make house calls, look after the illnesses of every member of the family, was even able to call the family dog by name. Whole new academic departments have been installed—some of them, in the state-run medical schools, actually legislated into existence—called, in the official catalogues, *Family Practice*, *Primary Health Care*, *Preventive Medicine*, *Primary Medicine*. The avowed intention is to turn out more general practitioners of the type that everyone remembers from childhood or from one's parents' or grandparents' childhood, or from books, movies, and television.

What is it that people have always expected from the doctor? How, indeed, has the profession of medicine survived for so much of human history? Doctors as a class have always been criticized for their deficiencies. Montaigne in his time, Molière in his, and Shaw had less regard for doctors and their medicine than today's critics. What on earth were the patients of physicians in the nineteenth century and the centuries before, all the way back to my professional ancestors, the shamans of prehistory, hoping for when they called for the doctor? In the years of the great plagues, when carts came through the town streets each night to pick up the dead and carry them off for burial, what was the function of the doctor? Bubonic plague, typhus, tuberculosis, and syphilis were representative examples of a great number of rapidly progressive and usually lethal infections, killing off most of the victims no matter what was done by the doctor. What did the man do, when called out at night to visit the sick for whom he had nothing to offer for palliation, much less cure?

Well, one thing he did, early on in history, was plainly magic. The shaman learned his profession the hardest way: he was compelled to go through something like a version of death itself, personally, and when he emerged he was considered qualified to deal with patients. He had epileptic fits, saw visions, and heard voices, lost himself in the wilderness for weeks on end, fell into long stretches of coma, and when he came back to life he was licensed to practice, dancing around the bedside, making smoke, chanting incomprehensibilities, and *touching* the patient everywhere. The touching was the real professional secret, never acknowledged as the central, essential skill, always obscured by the dancing and the chanting, but always busily there, the laying on of hands.

There, I think, is the oldest and most effective act of doctors, the touching. Some people don't like being handled by others, but not, or almost never, sick people. They *need* being touched, and part of the dismay in being very sick is the lack of close human contact. Ordinary people, even close friends, even family members, tend to stay away from the very sick, touching them as infrequently as possible for fear of interfering, or catching the illness, or just for fear of bad luck. The doctor's oldest skill in trade was to place his hands on the patient.

Over the centuries, the skill became more specialized and refined, the hands learned other things to do beyond mere contact. They probed to feel the pulse at the wrist, the tip of the spleen, or the edge of the liver, thumped to elicit resonant or dull sounds over the lungs, spread ointments over the skin, nicked veins for bleeding, but the same time touched, caressed, and at the end held on to the patient's fingers.

Most of the men who practiced this laying on of hands must have possessed, to begin with, the gift of affection. There are, certainly, some people who do not like other people much, and they would have been likely to stay away from an occupation requiring touching. If, by mistake, they found themselves apprenticed for medicine, they probably backed off or, if not, turned into unsuccessful doctors.

Touching with the naked ear was one of the great advances in the history of medicine. Once it was learned that the heart and lungs made sounds of their own, and that the sounds were sometimes useful for diagnosis, physicians placed an ear over the heart, and over areas on the front and back of the chest, and listened. It is hard to imagine a friendlier human gesture, a more intimate signal of personal concern and affection, than these close bowed heads affixed to the skin. The stethoscope was invented in the nineteenth century, vastly enhancing the acoustics of the thorax, but removing the physician a certain distance from his patient. It was the earliest device of many still to come, one new technology after another, designed to increase that distance.

Today, the doctor can perform a great many of his most essential tasks from his office in another building without ever seeing the patient. There are even computer programs for the taking of a history: a clerk can ask the questions and check the boxes on a printed form, and the computer will instantly provide a printout of the diagnostic possibilities to be considered and the laboratory procedures to be undertaken. Instead of spending forty-five minutes listening to the chest and palpating the abdomen, the doctor can sign a slip which sends the patient off to the X-ray department for a CT scan, with the expectation of seeing within the hour, in exquisite detail, all the body's internal organs which he formerly had to make guesses about with his fingers and ears. The biochemistry laboratory eliminates the need for pondering and waiting for the appearance of new signs and symptoms. Computerized devices reveal electronic intimacies of the flawed heart or malfunctioning brain with a precision far beyond the touch or reach, or even the imagining, of the physician at the bedside a few generations back.

The doctor can set himself, if he likes, at a distance, remote from the patient and the family, never touching anyone beyond a perfunctory handshake as the first and only contact. Medicine is no longer the laying on of hands, it is more like the reading of signals from machines.

The mechanization of scientific medicine is here to stay. The new medicine works. It is a vastly more complicated profession, with more things to be done on short notice on which issues of life or death depend. The physician has the same obligations that he carried, overworked and often despairingly, fifty years ago, but now with any number of technological maneuvers to be undertaken quickly and with precision. It looks to the patient like a different experience from what his parents told him about, with something important left out. The doctor seems less like the close friend and confidant, less interested in him as a person, wholly concerned with treating the disease. And there is no changing this, no going back; nor, when you think about it, is there really any reason for wanting to go back. If I develop the signs and symptoms of malignant hypertension, or cancer of the colon, or subacute bacterial endocarditis, I want as much comfort and friendship as I can find at hand, but mostly I want to be treated quickly and effectively so as to survive, if that is possible. If I am in bed in a modern hospital, worrying about the cost of that bed as well, I want to get out as fast as possible, whole if possible.

In my father's time, talking with the patient was the biggest part of medicine, for it was almost all there was to do. The doctor–patient relationship was, for better or worse, a long conversation in which the patient was at the epicenter of concern and knew it. When I was an intern and scientific technology was in its earliest stage, the talk was still there, but hurried, often on the run.

Today, with the advance of medicine's various and complicated new technologies, the ward rounds now at the foot of the bed, the drawing of blood samples for automated assessment of every known (or suggested) biochemical abnormality, the rolling of wheelchairs and litters down through the corridors to the X-ray department, there is less time for talking. The longest and most personal conversations held with hospital patients when they come to the hospital are discussions of finances and insurance, engaged in by personnel trained in accountancy, whose scientific instruments are the computers. The hospitalized patient feels, for a time, like a working part of an immense, automated apparatus. He is admitted and discharged by batteries of computers, sometimes without even learning the doctors' names. The difference can be strange and vaguely dismaying for patients. But there is another difference, worth emphasis. Many patients go home speedily, in good health, cured of their diseases. In my father's day this happened much less often, and when it did, it was a matter of good luck or a strong constitution. When it happens today, it is more frequently due to technology.

There are costs to be faced. Not just money, the real and heavy dollar costs. The close-up, reassuring, warm touch of the physician, the comfort and concern, the long, leisurely discussions in which everything including the dog can be worked into the conversation, are disappearing from the practice of medicine, and this may turn out to be too great a loss for the doctor as well as for the patient. This uniquely subtle, personal relationship has roots that go back into the beginnings of medicine's history, and needs preserving. To do it right has never been easy; it takes the best of doctors, the best of friends. Once lost, even for as short a time as one generation, it may be too difficult a task to bring it back again.

If I were a medical student or an intern, just getting ready to begin, I would be more worried about this aspect of my future than anything else. I would be apprehensive that my real job, caring for sick people, might soon be taken away, leaving me with the quite different occupation of looking after machines. I would be trying to figure out ways to keep this from happening.

# 7
# NURSES

When my mother became a registered nurse at Roosevelt Hospital, in 1903, there was no question in anyone's mind about what nurses did as professionals. They did what the doctors ordered. The attending physician would arrive for his ward rounds in the early morning, and when he arrived at the ward office the head nurse would be waiting for him, ready to take his hat and coat, and his cane, and she would stand while he had his cup of tea before starting. Entering the ward, she would hold the door for him to go first, then his entourage of interns and medical students, then she followed. At each bedside, after he had conducted his examination and reviewed the patient's progress, he would tell the nurse what needed doing that day, and she would write it down on the part of the chart reserved for nursing notes. An hour or two later he would be gone from the ward, and the work of the rest of the day and the night to follow was the nurse's frenetic occupation. In addition to the stipulated orders, she had an endless list of routine things to do, all learned in her two years of nursing school: the beds had to be changed and made up with fresh sheets by an exact geometric design of folding and tucking impossible for anyone but a trained nurse; the patients had to be washed head to foot; bedpans had to be brought, used, emptied, and washed; temperatures had to be taken every four hours and meticulously recorded on the chart; enemas were to be given; urine and stool samples collected, labeled, and sent off to the laboratory; throughout the day and night, medications of all sorts, usually pills and various vegetable extracts and tinctures, had to be carried on trays from bed to bed. At most times of the year about half of the forty or so patients on the ward had typhoid fever, which meant that the nurse couldn't simply move from bed to bed in the performance of her duties; each typhoid case was screened from the other patients, and the nurse was required to put on a new gown and wash her hands in disinfectant before approaching the bedside. Patients with high fevers were sponged with cold alcohol at frequent intervals. The late-evening back rub was the rite of passage into sleep.

In addition to the routine, workaday schedule, the nurse was responsible for responding to all calls from the patients, and it was expected that she would do so on the run. Her rounds, scheduled as methodical progressions around the ward, were continually interrupted by these calls. It was up to her to evaluate each situation quickly: a sudden abdominal pain in a typhoid patient might signify intestinal perforation; the abrupt onset of weakness, thirst, and pallor meant intestinal hemorrhage; the coughing up of gross blood by a tuberculous patient was an emergency. Some of the calls came from neighboring patients on the way to recovery; patients on open wards always kept a close eye on each other: the man in the next bed might slip into coma or seem to be dying, or be indeed dead. For such emergencies the nurse had to get word immediately to the doctor on call, usually the intern assigned to the ward, who might be off in the outpatient department or working in the diagnostic laboratory (interns of that day did all the laboratory work themselves; technicians had not yet been invented) or in his room. Nurses were not allowed to give injections or to do such emergency procedures as spinal punctures or chest taps, but they were expected to know when such maneuvers were indicated and to be ready with appropriate trays of instruments when the intern arrived on the ward.

It was an exhausting business, but by my mother's accounts it was the most satisfying and rewarding kind of work. As a nurse she was a low person in the professional hierarchy, always running from place to place on orders from the doctors, subject as well to strict discipline from her own administrative superiors on the nursing staff, but none of this came through in her recollections. What she remembered was her usefulness.

Whenever my father talked to me about nurses and their work, he spoke with high regard for them as professionals. Although it was clear in his view that the task of the nurses was to do what the doctor told them to, it was also clear that he admired them for being able to do a lot of things he couldn't possibly do, had never been trained to do. On his own rounds later on, when he became an attending physician himself, he consulted the ward nurse for her opinion about problem cases and paid careful attention to her observations and chart notes. In his own days of intern training (perhaps partly under my mother's strong influence, I don't know) he developed a deep and lasting respect for the whole nursing profession.

I have spent all of my professional career in close association with, and close dependency on, nurses, and like many of my faculty colleagues, I've done a lot of worrying about the relationship between medicine and nursing. During most of this century the nursing profession has been having a hard time of it. It has been largely, although not entirely, an occupation for women, and sensitive issues of professional status, complicated by the special issue of the changing role of women in modern society, have led to a standoffish, often adversarial relationship between nurses and doctors. Already swamped by an increasing load of routine duties, nurses have been obliged to take on more and more purely administrative tasks: keeping the records in order; making sure the supplies are on hand for every sort of ward emergency; supervising the activities of the new paraprofessional group called LPNs (licensed practical nurses), who now perform much of the bedside work once done by RNs (registered nurses); overseeing ward maids, porters, and cleaners; seeing to it that patients scheduled for X rays are on their way to the X-ray department on time. Therefore, they have to spend more of their time at desks in the ward office and less time at the bedsides. Too late maybe, the nurses have begun to realize that they are gradually being excluded from the one duty which had previously been their most important reward but which had been so taken for granted that nobody mentioned it in listing the duties of a nurse: close personal contact with patients. Along with everything else nurses did in the long day's work, making up for all the tough and sometimes demeaning jobs assigned to them, they had the matchless opportunity to be useful friends to great numbers of human beings in trouble. They listened to their patients all day long and through the night, they gave comfort and reassurance to the patients and their families, they got to know them as friends, they were depended on. To contemplate the loss of this part of their work has been the deepest worry for nurses at large, and for the faculties responsible for the curricula of the nation's new and expanding nursing schools. The issue lies at the center of the running argument between medical school and nursing school administrators, but it is never clearly stated. Nursing education has been upgraded in recent years. Almost all the former hospital schools, which took in high-school graduates and provided an RN certificate after two or three years, have been replaced by schools attached to colleges and universities, with a four-year curriculum leading simultaneously to a bachelor's degree and an RN certificate.

The doctors worry that nurses are trying to move away from their historical responsibilities to medicine (meaning, really, to the doctors' orders). The nurses assert that they are their own profession, responsible for their own standards, coequal colleagues with physicians, and they do not wish to become mere ward administrators or technicians (although some of them, carrying the new and prestigious title of "nurse practitioner," are being trained within nursing schools to perform some of the most complex technological responsibilities in hospital emergency rooms and intensive care units). The doctors claim that what the nurses really want is to become substitute psychiatrists. The nurses reply that they have unavoidable responsibilities for the mental health and well-being of their patients, and that these are different from the doctors' tasks. Eventually the arguments will work themselves out, and some sort of agreement will be reached, but if it is to be settled intelligently, some way will have to be found to preserve and strengthen the traditional and highly personal nurse-patient relationship.

I have had a fair amount of firsthand experience with the issue, having been an apprehensive patient myself off and on over a three-year period on the wards of the hospital for which I work. I am one up on most of my physician friends because of this experience. I know some things they do not know about what nurses do.

One thing the nurses do is to hold the place together. It is an astonishment, which every patient feels from time to time, observing the affairs of a large, complex hospital from the vantage point of his bed, that the whole institution doesn't fly to pieces. A hospital operates by the constant interplay of powerful forces pulling away at each other in different directions, each force essential for getting necessary things done, but always at odds with each other. The intern staff is an almost irresistible force in itself, learning medicine by doing medicine, assuming all the responsibility within reach, pushing against an immovable attending and administrative staff, and frequently at odds with the nurses. The attending physicians are individual entrepreneurs trying to run small cottage industries at each bedside. The diagnostic laboratories are feudal fiefdoms, prospering from the insatiable demands for their services from the interns and residents. The medical students are all over the place, learning as best they can and complaining that they are not, as they believe they should be, at the epicenter of everyone's concern. Each individual worker in the place, from the chiefs of surgery to the dieticians to the ward maids, porters, and elevator operators, lives and works in the conviction that the whole apparatus would come to a standstill without his or her individual contribution, and in one sense or another each of them is right.

My discovery, as a patient first on the medical service and later in surgery, is that the institution is held together, *glued* together, enabled to function as an organism, by the nurses and by nobody else.

The nurses, the good ones anyway (and all the ones on my floor were good), make it their business to know everything that is going on. They spot errors before errors can be launched. They know everything written on the chart. Most important of all, they know their patients as unique human beings, and they soon get to know the close relatives and friends. Because of this knowledge, they are quick to sense apprehensions and act on them. The average sick person in a large hospital feels at risk of getting lost, with no identity left beyond a name and a string of numbers on a plastic wristband, in danger always of being whisked off on a litter to the wrong place to have the wrong procedure done, or worse still, *not* being whisked off at the right time. The attending physician or the house officer, on rounds and usually in a hurry, can murmur a few reassuring words on his way out the door, but it takes a confident, competent, and cheerful nurse, there all day long and in and out of the room on one chore or another through the night, to bolster one's confidence that the situation is indeed manageable and not about to get out of hand.

Knowing what I know, I am all for the nurses. If they are to continue their professional feud with the doctors, if they want their professional status enhanced and their pay increased, if they infuriate the doctors by their claims to be equal professionals, if they ask for the moon, I am on their side.

# 8
# NEUROLOGY

During my second year in medical school I took an elective course in "advanced" neuroanatomy, taught by Professor David Rioch, a member of the Department of Anatomy. The course consisted entirely of the construction, by each member of the class of a dozen or so students, of a plasticene model of the human brain. It was child's play, in several senses: building an engrossing sort of toy, made up of dowels fitted to each other and adorned with wire extensions upon which the various nuclear structures in the brain were molded in clay of various colors; putting together an esthetic experience so that it came out right without falling apart; and, in the end, regarding in puzzlement an altogether primitive and naïve conception of the immense and, at that time, unimaginable complexity of the brain. Even so, in 1934, when we had no idea that we had no idea, the finished model seemed to make a certain kind of sense. Our wires arranged themselves to represent the great sensory tracts entering the base of the brain from the spinal cord, intermeshed with the bundles of motor fibers on their way down from the brain stem, and, higher up, the fanning out of intercommunicating fibers linking the basal ganglia, the cerebellum, the thalamus, and some smaller globules of clay representing structures in the hypothalamus, all neatly wired to each other and out and up into the highest reaches of the cerebral cortex, from which, as our models were intended to demonstrate, everything else was governed and run, like a wonderful electric apparatus. When the models had been completed, and checked in detail by the professor, we sprayed the things with varnish and took them back to our rooms. Mine pleased me so much that I kept it for nearly fifteen years, until it finally dried up and fell to pieces on the back shelf of a closet.

This simple mechanical exercise, requiring only three months, provided lasting enchantment. During the next two years of school, while I was clerking in the various teaching hospitals of Boston, the patients that interested me the most, and on whom I spent the most time on the wards and back in the library, were cases of neurological disease.

The making of a neurological diagnosis was itself a kind of game. All you needed to play were three implements, a rubber hammer for eliciting reflexes over tendons and muscles, a pin for testing pain receptors, and a wisp of cotton (some neurologists carried around a feather) for testing light touch. One's thumbnail served for scratching the soles of the feet to see if the toes extended and spread; this reflex, the Babinski, was probably the most important single feature of the neurological examination, signifying that damage had been done to the long motor pathways and, always, real trouble.

I had it in the back of my mind, all through the last years of medical school and my internship in medicine, that one day I would study neurology seriously. Hopkins, McGill, Columbia, the University of Pennsylvania, and Harvard were the places to try for, I had been told, but I had no idea whether one was better than another, or whether they were good at different things. Then, one day in 1938, partway through my internship, I was told that Dr. Robert F. Loeb was taking on the directorship of the Neurological Institute of New York and that I should look into the possibility of a residency there. Loeb was a youngish but already famous member of the faculty in the Department of Medicine at P&S, recognized internationally for his work on Addison's disease, the metabolic functions of the adrenal cortex, and the new field of salt and water control in physiology. He had never done neurology, but had been persuaded to leave his position at Presbyterian Hospital, where he was a full professor, to reorganize and modernize the Neurological Institute and, especially, to try to introduce more contemporary research into what had always been a prestigious clinical facility but had had very limited scientific interests.

This was a piece of exciting news. Neurology had always been an entirely descriptive branch of medicine. Once the clinician had figured out the precise location of the lesion (or lesions) in the brain or spinal cord—and the exactitude with which this localization could be accomplished, when you knew enough neuroanatomy, was the challenge of the field—there was nothing much to be done for therapy because of so little understanding of how the structures really worked. The principal exceptions were pernicious anemia, often associated with destruction of the long sensory and motor pathways in the spinal cord, which could be treated with liver extract;

neurosyphilis, for which fever therapy was marginally useful; and pellagra, by this time a rare disease seen only in chronic alcoholics, in which widespread lesions of both the central and peripheral nervous systems occurred because of vitamin B deficiency. A few brain tumors could be successfully removed by the neurosurgeons, but the commonest ones were almost always inoperable. The great need in neurology in the 1930s was obviously to introduce some proper research laboratories devoted to the study of brain disease.

The Neurological Institute was dominated at that time by a large attending staff, almost all of whom were in private practice and carried the professional title of "neuropsychiatrist"; one couldn't make a living doing just neurology, so most of the clinicians did psychiatry as well, and many of the private rooms in the institute were occupied by "nervous" patients, not sick enough to be confined next door in the New York State Psychiatric Institute, but certainly without real neurological disease either. Some were unhappy neurotics, needing rest and comfort, others alcoholics in for drying out, and a few barbiturate addicts admitted for what usually turned out to be unsuccessful efforts at withdrawal.

The real neurological diseases were out on the open wards, twenty beds to a ward, and these beds were governed with considerable autonomy by the neurology residents, guided by a carefully selected group of attending neurologists who made ward rounds each morning with the residents, along with clusters of Columbia medical students on neurology elective clerkships. Once a week Dr. Loeb came down from his new office on the top floor of the institute to make his own grand rounds on one ward or another. These were state occasions, involving not only the neurologists but also a large entourage of residents, students, and visitors from the medical service across the street in Presbyterian Hospital. We spent a lot of time selecting the cases to be presented at these rounds; they had to be interesting combinations of complex neurological and medical illnesses, and we always hoped to have a few patients in whom the diagnostic problems would be sufficiently obscure to baffle Loeb himself, but this, so far as I can recall, never happened. Loeb was a master diagnostician. He had the gift I had observed earlier in Hermann Blumgart at the Beth Israel Hospital at Harvard: he could walk on a ward and recognize, by some kind of instinct, each of the patients in whom something deeply serious was going on. He was also a master at the art of raising interesting questions, unanswerable but nevertheless interesting, about the possible mechanisms that underlay the diseases with which we were confronted. What did we think was going on, *really* going on, in such a disease as multiple sclerosis (this was one of the commonest of all problems on our wards)? It begins in young adults with a series of sudden, small neurological defects: double vision, slurring speech, unsteadiness of gait, weakness in one extremity or another, areas of numbness here or there, then progresses in cyclic episodes of new and equally sudden impairments, all due to patches of destruction of the myelin sheaths around nerve fibers in one or another region of the brain and spinal cord. The disease goes on for years in most patients, but not all; some have only one or two brief attacks and then, unaccountably, are finished with the disease, while others go on to total incapacitation for the rest of their lives. It was during Loeb's rounds, and the long discussions resulting from his questions, that I became convinced that multiple sclerosis was an autoimmune disease, caused by the presence within the brain of antibodies directed against a component of brain tissue itself. It was a new idea in the 1930s. Thomas

Rivers, George Packer Berry, and Francis Schwentker had shown five years earlier at the Rockefeller Institute that monkeys developed brain lesions resembling multiple sclerosis after being injected repeatedly with extracts of monkey brain tissue; their observation had been an accidental one, made in the course of attempts to develop vaccines against various viruses known to cause brain disease. The problem in the ward rounds discussions was how to extrapolate from this experimental model to the spontaneous human disease; how to explain this disease, progressing with one destructive lesion after another over many years, in terms of an antibrain antibody? The matter remains almost as much an unsolved problem today; there are, however, new sources for clues: it is suspected that long-latent viruses, measles for example, may become lodged in brain tissue and later give rise to antibodies simultaneously directed at the virus and a component of brain tissue; it is also known that multiple sclerosis patients differ as a group from other people with respect to the HLA gene locus, which governs immunologic reactivity. Perhaps these patients have an inborn error of immunologic perception, which permits the elaboration of antiself antibodies in the brain in response to the presence of an otherwise irrelevant virus, and thus the disaster.

In 1940 Dr. Loeb went back to the Department of Medicine at P&S and soon thereafter became its chairman. Dr. Tracy Jackson Putnam arrived from Harvard to become director of the Neurological Institute. Putnam was primarily a neurosurgeon, but he had a long and distinguished record in research; the discovery for which he is still remembered today, along with his collaborator Houston Merritt, was the dilantin class of anticonvulsive drugs for the treatment of epilepsy.

Putnam promptly organized a laboratory for work on a method for producing brain abscesses in experimental animals, so that better methods could be devised for treating such abscesses by combining surgery with chemotherapy, using one or another of the new sulfonamide drugs. I began working on this during the summer of 1940, when the residents' duties on the wards were relatively light.

We devised a method that worked beautifully and with reproducibility, providing chronic brain abscesses that could be observed for many weeks and that resembled in detail the counterpart lesions in the brains of human beings, and I wrote my first scientific paper. Very soon thereafter, penicillin became available, then other more powerful antibiotics, and there was no need for an animal model for studying brain abscesses. The disease itself became a rarity, and still is.

I finished my residency and then became the first Tilney Fellow in Neurology at the institute, thus assured of an income of $1800 for the year 1941, with the understanding that I would spend the year at Harvard and then return to New York as chief resident at the Neurological Institute. I was twenty-seven years old, and for the first time in my life I was independent enough, and solvent enough, to undertake marriage. Beryl and I were married in the chapel of Grace Church in Manhattan on the morning of New Year's Day, 1941, and that afternoon we headed for Boston. We had an extremely small but absolutely perfect apartment on Longwood Avenue, just across the street from the Children's Hospital and down the road from the medical school quadrangle, and settled in after New Year's Day. A few days later we learned that John Dingle's laboratory was being mobilized for a trip to Halifax, Nova Scotia, where a meningitis epidemic had just been recognized and the health authorities, shorthanded because of the war, had requested help from Harvard. So we packed again and flew to Halifax, where I went to work on the treatment of meningococcal meningitis with a new sulfonamide called sulfadiazine, of which I had never heard, and Beryl was recruited as a laboratory assistant to keep the records and carry cultures from one place to another.

We were in Halifax for about a month, culturing the spinal fluids of several hundred patients with meningitis, collecting samples of serum from these patients and other people who did not develop meningitis but were in close contact, in order to study the possible role of antibodies in protection against the disease, and recording with care the clinical course of the illness under treatment with sulfadiazine, which was administered to all patients with an established diagnosis. Sulfadiazine was wonderfully effective. The only patients who failed to recover were those with a rapidly developing and overwhelming infection—some of them became comatose within a few hours and were brought to the hospital in deep shock, their skin surfaces covered everywhere by areas of hemorrhagic necrosis (looking very much like the Shwartzman phenomenon which I was to study several years later), and these patients were dead before we could start treatment. All the rest, the majority, recovered promptly when given sulfadiazine, and we saw none of the late complications—blindness, deafness, mental confusion—which had occurred in earlier epidemics of untreated meningococcal meningitis.

We came back to Boston with crates of cultures and sera, and my laboratory was committed to the problem of the meningococcus and the mechanism of its peculiar affinity for the surfaces of the brain and spinal cord in human beings. None of the conventional laboratory animals were particularly vulnerable to this organism: rabbits, guinea pigs, rats, and mice could tolerate the intravenous injection of huge numbers of live meningococci without turning a hair, and the bacteria disappeared from their bloodstreams within ten minutes or so. It was evident that the animals possessed a highly effective mechanism for their protection, and I settled down to find out more about this. The first and simplest possibility, that they were able to kill off the injected meningococci by means of an already-existing "natural" antibody, was easiest to test in rabbits, so rabbits became the laboratory's routine animal. We quickly learned that the serum of a normal adult rabbit was capable of destroying almost any number of meningococci; when up to a million organisms were added to a single milliliter of freshly obtained rabbit serum, and the mixture then incubated for a few hours at 37 degrees Centigrade, the specimens became sterile. If the serum samples were heated at 56 degrees Centigrade for an hour before adding the bacteria, the bactericidal action was completely lost, indicating that the killing power depended on the presence of complement (a sequence of proteins, still incompletely understood, which makes possible the action of antibodies against antigens on the surface of bacteria).

We thought it useful, given so powerful an example of natural immunity already in existence in animals, to see whether we could obtain even stronger antibacterial sera by immunizing the rabbits. We injected animals with suspensions of heat-killed meningococci, and collected sera at weekly intervals. These samples were set up as in the initial experiments, adding various numbers of live bacteria to the serum specimens and determining how many were killed, and how quickly. Within the next few days we encountered our paradox: the sera from the immunized rabbits, which had been capable of killing a million meningococci in a few hours, had now lost this property. There were potent and specific antibodies in these sera, as we could show in other kinds of tests—agglutination, precipitation, and complement fixation tests. But, with the appearance of a specific antibody, the bactericidal activity vanished.

Moreover, something of the same sort could be shown in the whole rabbit, *in vivo*. When we injected live bacteria into the bloodstream of our immunized animals, and then measured the survival of the bacteria by serial blood cultures, we were surprised to learn that the blood cultures were still positive twenty-four hours later in the more intensively immunized rabbits, in contrast to the unimmunized animals, in which all of the meningococci had disappeared within ten to fifteen minutes.

By this time it was late April of 1941 and I was in a hurry. The problem had turned into something fascinating, involving both paradox and surprise. I knew I was expected back in New York the next January to become a neurologist, so I worked as fast as I could. What I had run into was an antique immunologic phenomenon called the "prozone," in which an excess of antibody turns off the immune reaction unless the serum is sufficiently diluted. However, the difference in my laboratory—what was new—was that it worked *in vivo*: an immunized animal could lose, as the result of being immunized, its own natural defense. This might, I thought, have useful implications for susceptibility in certain human infections beyond meningitis—typhoid fever and brucellosis, for example—and I wanted to get on with it.

However, as it turned out, I never got to finish the problem or even answer the principal questions. Nor did I ever get back to the Neurological Institute. The Rockefeller Institute was put on notice in late 1941, then mobilized as a naval medical research unit; I was assigned to it as lieutenant, and received orders to turn up in New York, in uniform, by the end of March 1942. John Dingle and I reluctantly agreed to bring the still inconclusive problem of the *in vivo* prozone to a premature end and write the work up; to this day I've never been able to return, full-time, to the problem. It still hangs there in my mind, and I don't believe any other laboratory has ever settled it.

I think I would still pick neurology as the most fascinating of all fields in medicine. It is now beginning to move into problems originally staked out by psychiatry, and the contributions from neuropharmacology have already begun to transform the discipline. The most enthralling of these is endorphin, a simple peptide secreted within the brain with the particular function of attaching specifically to the surface of cells responsible for the awareness of pain, at the same sites to which morphine and heroin habitually become attached.

These things are interesting for all kinds of reasons, some of them urgently important. Now that the chemical structures are known, it may be possible to design new classes of drugs for pain, perhaps without the side effects and addicting properties of morphine. It is also conceivable that new insights can be gained into the mechanism of addiction itself, and perhaps new ways will be found to cope with at least the purely medical aspects of one of this century's most appalling social problems. Perhaps, as well, when we have learned enough about endorphin, and gotten used to the idea that such a thing exists in our brains, we may take a different attitude toward addiction. I wonder what would happen if pharmacologic science were to produce a "natural" drug, as natural, say, as endorphin, possessing the subjectively pleasurable properties of heroin, but without addiction. Would it be allowed, or would we pass laws to forbid it?

But the most interesting question of all is why does such a substance exist? What is the biological purpose of endorphin? Is its antipain function what it is really designed to accomplish, or is this a more or less incidental side effect, a biological accident, with some other still-unguessed-at role in the regulation of messages in the brain?

If it is, in fact, a built-in mechanism for the alleviation of pain, how did it get there past all the selective tests of evolution? Why should it have survival value for a species, or for an individual animal? For this is what you would have to find, unless its existence is to make no sense in the context of modern biology. We take it for granted that every major inherited trait possessed uniformly by any species is there because of natural selection. This is as solid and inflexible a rule as any in science.

It would be a different problem if only we humans made endorphin in our brains. Perhaps you could make the case that for a species as intelligent and at the same time as interdependent and watchful of each other as ours, it might be useful to install a device of this kind to guard against intolerable pain, or to ease the individual through what might otherwise be an agonizing process of dying. Without it, living in our kind of intimacy, at our close quarters, might be too difficult for us, and we might separate from one another, each trying life on his own, and the species would then, of course, collapse.

But why should mice have the same equipment, and every other vertebrate thus far studied?

And why, of all creatures, earthworms? For it has just been

discovered that the primitive nervous system of annelids is richly endowed with the same kind of endorphin receptors, and it can be assumed that the worm possesses the same system for pain relief as exists in our own brains. I am glad to learn of this. Earthworms do have sensory equipment, I know. They withdraw quickly when touched, even when blown upon. Without protection against overwhelming pain, the day-to-day life of a worm, being stepped on, snatched by birds, ground under plows, washed away in streams, would be hellish indeed.

Perhaps this is simply a piece of extraordinary good luck on the part of nature. Maybe something slipped up somewhere early in evolution, and all of us were endowed with something ineffable, free for the having, carrying no particular value for competition. The genes were simply handed down, species after elaborate species, to restrain the suffering of living and dying, by pure chance. I have to doubt this, as an earnest believer in the details of evolution.

Yet there it is, a biologically universal act of mercy. I cannot explain it, except to say that I would have put it in had I been around at the very beginning, sitting as a member of a planning committee, say, and charged with the responsibility for organizing for the future a closed ecosystem crowded with an infinite variety of life on this planet. No such system could possibly operate without pain, and pain receptors would have to be planned in detail for all sentient forms of life, plainly for their own protection and the avoidance of danger. But not limitless pain; this would have the effect of turbulence, unhinging the whole system in an agony even before it got under way. And not, I should think, the awareness of dying. I would have cast a vote for a modulator of pain, finely enough adjusted to assure its usefulness, but set with a governor of some sort, to make sure it never could get out of hand. In this sense, endorphin may have developed in our brains not for its selective value to our species, or any species, or any individuals without species, but for the survival and perpetuation of the whole biosphere, or as it is sometimes called, the System.

No one can predict how the endorphin story will turn out in the end, for it is only at its beginning. At the present stage, it might go anywhere, mean anything. Conceivably, chemical messengers of this class, small peptide molecules, may even be involved in disorders of the brain, including schizophrenia.

This state of affairs tells a central truth about research. Making guesses at what might lie ahead, when the new facts have arrived, is the workaday business of science, but it is never the precise, surefooted enterprise that it sometimes claims credit for being. Accurate prediction is the accepted measure of successful research, the ultimate reward for the investigator, and also for his sponsors. Convention has it that prediction comes in two sequential epiphanies: first, the scientist predicts that his experiment will turn out the way he predicts; and then, the work done, he predicts what the experiment says about future experiments, his or someone else's. It has the sound of an intellectually flawless acrobatic act. The mind stands still for a moment, leaps out into midair at precisely the millisecond when a trapeze from the other side is hanging at the extremity of its arc, zips down, out, and up again, lets go and flies into a triple somersault, then catches a second trapeze timed for that moment and floats to a platform amid deafening applause. There is no margin for error. Success depends not so much on the eye or the grasp, certainly not on the imagination, only on the predictable certainty of the release of the bars to be caught. Clockwork.

It doesn't actually work this way, and if scientists thought it did, nothing would get done; there would be only a mound of bone-shattered scholars being carried off on stretchers.

In real life, research is dependent on the human capacity for making predictions that are wrong, and on the even more human gift for bouncing back to try again. This is the way the work goes. The predictions, especially the really important ones that turn out, from time to time, to be correct, are pure guesses. Error is the mode.

We all know this in our bones, whether engaged in science or in the ordinary business of life. More often than not, our firmest predictions are chancy, based on what we imagine to be probability rather than certainty, and we become used to blundering very early in life. Indeed, the universal experience, mandated in the development of every young child, of stumbling, dropping things, saying the words wrong, spilling oatmeal, and sticking one's thumb in one's eye are part of the preparation for adult living. A successful child is one who has learned so thoroughly about his own fallibility that he can never forget it, all the rest of his life.

In research, the usefulness of error is that it leads to more research, and this is what the word tells us. To *err* doesn't really mean getting things wrong; its etymology derives from the Indo-European root *ers*, signifying simply "to be in motion"; it comes into Latin as *errare*, meaning "to wander," but the same root emerges in Old Norse as *ras*, rushing about looking for something, from which we get the English word *race*. In order to get anything right, we are obliged first to get a great many things wrong.

The technical term *stochastic* is another word filled with the same lesson. We use it today to signify absolute randomness, and certain computers are programmed to turn out strings of stochastic variables in order that biomathematicians can arrange the appropriate controls for experiments involving large numbers of numbers. Stochastic is the jargon term for pure chance.

But it started out, as happens so often in language, with precisely the opposite meaning. The original Greek root was *stokhos*, meaning a brick column used as a target; from this the root words meaning "to take aim" were derived.

We like to think that we take aim and hit targets by taking advantage of a human gift for accuracy and precision. But there is this secret, embedded in the language itself: we become accurate only by trial and error, we tend to wander about, searching for targets. It is being in motion, at random (from a root meaning *running*, by the way), that permits us to get things done.

The immunologic system works in this way. When you inject a foreign antigen, horse serum protein, say, into a rabbit, a few lymphocytes are able to recognize this particular protein. They promptly begin to manufacture specific antibodies against the horse serum protein, and other cells of the same line begin dividing rapidly so that small factories for this kind of antibody production (and only this kind) are set up in the lymph nodes. The animal is now sensitized or immune, and stays that way indefinitely. When this phenomenon was first revealed, it was thought that the lymphocytes confronted by the horse serum molecules were somehow taught what to do by the encounter. Each cell, naïve to begin with, was *instructed* by the presence of the antigen, then learned how to make exactly the right antibody needed to lock precisely on to the foreign protein.

This notion, the "instructive" theory, reasonable as it

sounds, turned out to be wrong. It has now been replaced by what is called the "clonal selection" theory of immune response, supported by an immense body of solid research. According to this theory, lymphocytes are born knowing what to look for, and the individual cells, each with its individual kind of genetically determined receptor, roam the blood and tissues looking for the specific antigens which match the available specific receptors. When a lymphocyte meets its matching antigen, it promptly enlarges and begins dividing into identical progeny, all possessing the same receptors, and the result is a clone of identical cells all prepared to synthesize just the particular antibody needed, now and in the future. It is a tissue of cellular memory.

Among the billions of lymphocytes made available in a young animal are individual cells capable of recognizing the molecular configuration of almost anything in nature, including totally new, synthetic compounds never before seen in nature. The populations of such knowledgeable cells, and the extent of their collective repertoire, are vastly increased as the animal matures, probably as the result of somatic mutations or rearrangements of genes occurring from time to time in the stem cells which give rise to lymphocytes. The system works, and works with astonishing efficiency, because of the high mobility of the recognizing cells, their large numbers, and their capacity to amplify the antibody production quickly by replicating just the informed cells that are needed for the occasion.

It is eminently efficient, but from the point of view of any individual lymphocyte it must look like nothing but one mistake after another. When the horse serum protein appears, it is not recognizable to any but a small minority of the cell population; for all the rest it is a waste of time, motion, and effort. Also, there are risks all around, chances of making major blunders, endangering the whole organism. Flawed lymphocytes can turn up with an inability to distinguish between self and nonself, and replication of these can bring down the entire structure with the devastating diseases of autoimmunity. Blind spots can exist, or gaps in recognition analogous to color blindness, so that certain strains of animals are genetically unable to recognize the foreignness of certain bacteria and viruses.

Nevertheless, on balance the immune system works very well, so well indeed that the neurobiologists are currently entertaining (and being entertained by) the same selection theory to explain how the brain works. It is postulated that the thinking units equivalent to lymphocytes are the tiny columns of packed neurones which make up most of the substance of the cerebral cortex. These clusters are the receptors, prepared in advance for confrontation with this or that sensory stimulus, or this or that particular idea. For all the things we will ever see in the universe, including things not yet thought of, the human brain possesses one or another prepared, aware, knowledgeable cluster of connected neurones, as ready to lock on to that one idea as a frog's brain is for the movement of a fly. The recognition is amplified by synaptic alterations within the column of cells and among the other groups with which the column is connected, and memory is installed.

Statistically, the probability that any theory like this one, very early in its development, will turn out to be correct is of course vanishingly small, even with the speculative backing of an analogous mechanism in the immune system. The great thing about it, right or wrong, is that it is already causing ripples of interest and excitement, and other investigators are starting to plan experiments, cooking up ideas, their minds wandering, their receptors displayed at full attention, waiting for the *right* idea to come along. Neurology and immunology may be on the verge of converging.

I wrote a couple of essays a few years back on computers, in which I had a few things to say in opposition to the idea that machines could be made with what the computer people themselves call Artificial Intelligence; they always use capital letters for this technology, and refer to it in their technical papers as AI. I was not fond of the idea and said so, and proceeded to point out the necessity for error in the working of the human mind, which I thought made it different from the computer. In response, I received a great deal of mail, most of it gently remonstrative, but friendly, the worst kind of mail to get on days when things aren't going well anyway, pointing out to me in the simplest language how wrong indeed I was. Computers do proceed, of course, by the method of trial and error. The whole technology is based on this, can work in no other way.

One of the things I have always disliked about computers is that they are personally humiliating. They do resemble, despite my wish for it to be otherwise, the operations of the human mind. There are differences, but the Artificial Intelligence people, with their vast and clever computers, have come far enough along to make it clear that the machines behave like thinking machines. If they are right, the thing to worry about is not that they will ultimately be making electronic minds superior to ours but that already ours are so inferior to theirs, mine anyway. I have never heard of a computer, even a simple one, as dedicated to the deliberate process of forgetting information, losing it, restoring it out of context and in misleading forms, or generating such a condition of diffuse, inaccurate confusion as occurs every day in the average human brain. We are already so outclassed as to live in constant embarrassment.

I have been inputting, as they say, one bit of hard data after another into my brain all my life, some of it thruputting and outputting from the other ear, but a great deal held and stored somewhere, or so I am assured, but I possess no reliable device, anywhere in my circuitry, for retrieving it when needed. If I wish for the simplest of things, someone's name for example, I cannot send in a straightforward demand with any sure hope of getting back the right name. I am often required to think about something else, something quite unrelated, and wait, hanging around in the mind's lobby, picking up books and laying them down, pacing around, and then, if it is a lucky day, out pops the name. No computer could be designed by any engineer to function, or malfunction, in this way.

I have learned, one time or another, all sorts of things that I remember learning, but now they are lost to me. I cannot place the Thirty Years War or the Hundred Years War in the right centuries, nor have I at hand the barest facts about the issues involved. I once knew Keats, lots of Keats, by heart; he is still there, I suppose, probably scattered across the lobes of my left hemisphere, or maybe translated into the wordless language of my right hemisphere and preserved there forever as a set of hunches, but irretrievable as language. I have lost most of the philosophers I studied long ago and liked; the only sure memory I retain of Heidegger, even when I reread him today, is bewilderment. I have forgotten how to do cube roots, and will never learn again. Slide rules. Solid geometry. Thomas Hardy. Chinese etymology, which I learned in great volumes just a few years ago. The Bible, most of all the Sunday-school Bible, long since gone, obliterated.

It occurs to me that the computer-brain analogy needs to take account of what must otherwise seem an unnatural degree of fallibility on the part of the brain. Maybe what we do, by compulsion, in order to make sure that our minds are always reasonably well prepared to get us through any new day, is something like what happens to a computer when you walk past it carrying a powerful magnet. Perhaps we are in possession of similar devices—maybe chemical messengers of some sort—that periodically sweep the mind clear of surplus information, leaving the chips and circuits open to the new needs of the day. I cannot remember Keats because he was expunged one day; if I want him back, which I don't very badly, I am obliged to learn him all over again; he is gone out of my temporal lobe, where I had him once lodged.

In a way, this could be a reassuring notion, especially for anyone getting on, as I am, in years. It would be nice to know that I have a mechanism, even if it is beyond my control, that sweeps through my brain periodically, editing away the accumulations of old and no longer usable information, clearing the desk so to speak, disposing of all the old magazines and partly read books, getting the rooms of the mind ready for new lodgers. Indeed, if there were not such a mechanism, the brain would sooner or later be stuffed, swollen, bulging with facts, and unable to take in anything new. Signs would have to be displayed in all the lobes, reading OCCUPIED. Or NO ENTRY. Or, worst of all, signs repainted, changed to read EXIT.

Come to think of it, you could not run a human brain in any other way, and the clearing out of excess information must be going on, automatically, autonomically, all the time. Perhaps there are certain pieces of thought that must be classed as nonbiodegradable, like addition and one's family's names and how to read a taximeter, but a great deal of material is surely disposable. And the need for a quick and ready sanitation system is real: you cannot ever be sure, from minute to minute, when you will have to find a place to put something new. At the very least, you are required to have, and use, a mechanism for edging facts to one side, pushing them out of the way into something like a plastic kitchen bag. Otherwise you would run the risk of losing all good ideas. Have you noticed how often it happens that a really good idea—the kind of idea that looks, as it approaches, like the explanation for everything about everything—tends to hover near at hand when you are thinking hard about something quite different? On good days it happens all the time. There you are, halfway into a taxi, thinking hard about the condition of the cartilage in the right knee joint, and suddenly, with a whirring sound, in flies a new notion looking for a place to light. You'd better be sure you have a few bare spots, denuded of anything like thought, ready for its perching, or it will fly away into the dark. Computers cannot do this sort of thing. They can perform feats of mathematics beyond my comprehension, construct animated graphs at the touch of a finger, write with ease something like second-rate poetry, and they can even generate surprise for the operator, but I doubt very much that a computer, no matter how large and intricate, can itself *be* surprised, *feel* surprised; there isn't room enough for that.

Computers are good at seeing patterns, better than we are. They can connect things that seem unrelated to each other, scanning the night sky or the stained blotches of 50,000 proteins on an electrophoretic gel or the numbers generated by all the world's stock markets, and find relationships that matter. We do something like this with our brains, but we do it differently; we get things wrong. We use information not so much for its own sake as for leading to thoughts that really are unrelated, unconnected, patternless, and sometimes therefore quite new. If the human brain had not possessed this special gift, we would still be sharpening bones, muttering to ourselves, unable to make a poem or even whistle.

These two gifts, the ability to lose information unpredictably and to get relationships wrong, distinguish our brains from any computer I can imagine ever being manufactured. Artificial Intelligence is one thing, and I never spend a day without admiring it, but human intelligence is something else again. If I succeed in understanding the Shwartzman phenomenon, or in learning Homeric Greek, it will not be because the impulse to do so came in linear fashion from some prior stimulus. I will have blundered into it, thinking that something else led to it, when in fact the something else was heading in another direction, intent on other business.

This is not to say that I do not respect my mind, or anyone else's mind. I do, and I count it an added mark of respect to acknowledge that I do not understand it. My own mind, fallible, error-prone, forgetful, unpredictable, and ungovernable, is way over my head.

# 9
# GUAM AND OKINAWA

I had the guiltiest of wars, doing under orders one thing after another that I liked doing. The Rockefeller Institute (now University) had been signed up as a navy medical research unit, and I was taken on to work in the laboratory of the Rockefeller Hospital director, Dr. Thomas M. Rivers, who had one of the most extensive collections of viruses in the world stored under dry ice.

The Rockefeller Institute Hospital, mobilized as a naval unit in 1942, went to work in New York on several disease problems that were causing concern in the armed forces at that time: streptococcal infections and rheumatic fever, an epidemic form of pneumonia then known as primary atypical pneumonia ("atypical" because of the absence of pneumococci or other recognizable pathogens), hepatitis, yellow fever, malaria, parasitic infections, and meningitis. The hospital wards were filled with navy and marine personnel sent in from local training bases, most of them with pneumonia. My quarters were in Rivers's laboratory, which up to then had been devoted entirely to the study of highly pathogenic and uncongenial viruses; I had for my own use a dry-ice box filled with frozen samples of most of the known virus species, including rabies, equine encephalitis, rift valley fever, choriomeningitis, ten different varieties of psittacosis, and several samples of typhus rickettsiae (not a virus but classed with the viruses because of its inability to grow unless contained in living cells). Scrub typhus, a highly lethal disease also known as Tsutsugamushi fever, was of interest to the navy because of its occurrence in Japanese-occupied areas in the Pacific.

I was given four simultaneous assignments in the New York laboratory. The first was to try to isolate a virus from the patients with primary atypical pneumonia. The second was to learn whatever I could about typhus fever, in particular scrub typhus, and specifically to learn how to handle this agent without catching it; there had already been several deaths among laboratory workers in other institutions. The third was to continue some work on psittacosis among New York City pigeons, which had been in progress under Joseph Smadel in Rivers's laboratory for several years; this was not really a military problem but was thought to be of considerable public health importance for the city.

My fourth assignment was the only one in which I felt safe and unthreatened. Once a month I received a box filled with urine specimens from each of the naval bases in the Northeastern states, for Ascheim-Zondek tests. Pregnancy was something the navy worried a lot about, and the top officials had decided that Dr. Rivers's laboratory could be relied on for the appropriate facilities (mice, really) and also for confidentiality. Each month I reported two or three positive tests, resulting presumably in a return to civilian life of that number of Waves.

I spent most of that summer in Washington as a visiting investigator in Norman Topping's laboratory at the National Institutes of Health. Topping was the ranking authority on typhus fever, and his laboratory was busy with efforts to develop an effective vaccine against scrub typhus. I was taught how to cultivate the agent in chick embryos, how to prepare concentrated suspensions of the live agent for serological tests, and, above all, how to keep it from contaminating the air of the laboratory. The Waring blender had just been introduced as a useful instrument for homogenizing various tissues in research laboratories, and was in routine use in Topping's laboratory for getting smooth suspensions of the scrub typhus organism. It was not until later that year, long after I'd left, that it was recognized that taking the cover off the blender too soon after homogenization could be hazardous; one of the senior staff members, who had been my personal teacher in the summer, died of scrub typhus picked up from the air around the Waring blender.

In 1943 we were told that the Rockefeller naval research unit was to be sent to the Pacific, and preparations for the move were begun. It took a year to get ready. By the autumn of 1944 the purchases of laboratory equipment had been completed, and the necessary additions to the professional staff of the unit had been negotiated. When we left for the West Coast on the transcontinental railroad, we had representatives of almost every biological discipline in uniform and on board: entomologists, mammalogists, malacologists (snails), ornithologists, biochemists, clinicians, microbiologists, immunologists, virologists, two rickettsiologists (Jerome Syverton and me), and a large number of administrative officers and noncommissioned ranks who had been signed on as technicians. We ended up at San Bruno, just south of San Francisco, for a mandatory period of military training before shipping out.

I think we must have been one of the strangest lots of trainees ever to pass through San Bruno. We were all housed together in our own barracks, away from the thousands of other navy and marine people on their way to the Pacific. Early each morning we were assembled for drill, marching to the cadence of a full-throated marine sergeant who had little use for us; what he knew for sure about us was that we would be of little value in any hand-to-hand fight. Before drill, he called the roll, and it always went the same way, down through the alphabet; each of us was supposed to bark "Here!" the instant our name was called. Just before my name was Jerome Syverton, an extremely dignified professor from Rochester. Syverton's name was pronounced *Siv*erton, but the first day of drill the sergeant called out *Syv*erton with the long *y*; Syverton's response was "*Siv*erton—here!" This went on for four weeks—every morning the sergeant getting down to the *S*'s, all of us waiting, then the firmly and loudly mispronounced name, then Syverton's correction followed by "Here!" The moment had everything: the military versus the academic mind, a touch of class warfare, territoriality, human rights, pure fun. It lightened each day at its beginning.

Three days each week we were piled onto half-track trucks and driven west to the cliff overlooking the Pacific and taught to use the Browning Automatic Rifle—the BAR—the most terrifying and lethal weapon any of us had ever seen. One man in our group, an insect systematist and one of the country's major figures in taxonomy, but not quick on his feet and possessed of extremely thick eyeglass lenses, took up the instrument when it was his turn to fire and, his attention caught by something moving in the grass alongside, swung around in the direction of the drill sergeant with his finger on the trigger, almost but not quite firing, then waking up in surprise to see that everyone, the drill master and all his academic colleagues, were face down on the ground, yelling at him to put the damn thing down or point it toward the ocean.

We shipped out in early December, landing at Honolulu for a week or so, and then went on to Guam, a cluster of scientists and technicians in the midst of a huge convoy of marines and army infantry on their way to combat. We set up camp, living in tents while the construction work proceeded, and all of us—all ranks, from commander to marine private—built as fine a set of research laboratories as ever existed on a Pacific Island. The metal panels for the roofs and sides were labeled and ready to be screwed together; it was the simplest task to fit them together, hard to get wrong, but we finished with a sense of immense pride in our workmanship while we watched a crew of Seabees, professional construction workers, on a neighboring plot of ground putting up the island's navy hospital using the same materials. Later we learned that the Seabees had originally been scheduled to build our laboratories, but our commanding officer had decided that it would be better if we did it ourselves, thus preserving our political independence from other parts of the Guam establishment. This was believed to be important for the unit's future; we were supposed to be autonomous, capable of making our own selection of research problems to work on, not subject to orders from other authorities on the island who might, it was feared, be inclined to make a routine health department laboratory out of the unit.

The first call on our services came from Iwo Jima, where it was reported that tiny red mites had been encountered in the deep caves being taken from the Japanese. The invasion of that island had just started, and the casualities were beginning to arrive at the marine hospital down the road, along with rumors that the invasion was becoming an impossibility. The presence of mites added to the disaster; an outbreak of scrub typhus on Iwo Jima would have been a catastrophe. A team of technicians were sent to obtain samples of the insects, returning with the news that they were the wrong sort of mites, not the vectors of Tsutsugamushi fever. Congratulations were exchanged all round. The unit, which was always at risk for seeming an academic frivolity, a Guam ivory tower, was acknowledged for its usefulness to the navy, and our reputation was, at least for the time, solid.

At about the same time an outbreak of infectious hepatitis was reported from the Philippines, and a small group under Dr. George Mirick was flown off to pick up blood and fecal samples for study in the Guam laboratories. The disease was believed to be caused by a virus, but there was no information about the nature of the agent nor any experimental animal to which it could be transmitted. Beyond isolating the new cases and carrying out conventional epidemiological surveys in attempts to locate a source of the infection, there was nothing to be done about the problem. Nevertheless, it was a reassurance to the navy to know that the Guam unit was available and working on the matter.

Richard Shope occupied the laboratory across the path from mine, and in the early months on Guam there were no direct assignments for his group. He was at that time one of the country's leading virologists, having discovered the viruses of swine influenza and rabbit papilloma. He had already been told that he would be leading a team scheduled to land in Okinawa when the invasion took place the following spring, but he now had time on his hands. I watched him with fascination every morning, stepping out the front door of his laboratory holding a petrie dish containing blood agar in each hand. He performed the same ritual each day, arms outspread, holding the open dishes first to the north for a full minute, then in successive right turns, facing each compass point for whatever might land on his agar plates, then back inside to put the dishes in his incubator. He was looking for any wild microorganism which might possess antibiotic properties for the influenza virus being propagated in his mouse colony. Shope was not only persistent and careful, he had always been an extremely lucky investigator. Within a matter of weeks he had isolated a strain of *penicillium* with exactly the action he wanted: filtrates of broth cultures of the bug protected his mice against virus infections. It was, so far as I know, the first demonstration of an antiviral antibiotic. He named it, for his wife, Helen, "Helenin." It was effective against equine encephalitis viruses, not effective enough to be clinically useful, but an important clue for future work on antiviral agents.

While this was going on, Shope was organizing the contingent for Okinawa, ten officers and about twenty enlisted men, equipped with mobile laboratories for work on scrub typhus, insect and rodent studies, viruses, and a parasitology group with firsthand knowledge in schistosomiasis and the snails that transmit this disease. My assignment was to look for scrub typhus, and I had charge of a box containing fifty white mice for the purpose.

We left Guam for Okinawa on an army transport vessel in March and cast anchor in the harbor just north of the town of Naha two days after D day. The battle for the southern end of the island had already begun, and we could hear the bombing from our carrier aircraft and the cannon and small-arms fire and, at night, see the flashes a mile or so inland. It looked like a dangerous place, but we were more anxious to land than to stay on board the transport, which seemed to us a lot more hazardous. The kamikaze raids had begun, and we were sitting targets, along with scores of other ships densely packed in the harbor. Several ships were hit in the first days of the battle, but not ours. After a few days, I forget how many, we went ashore, climbing down a rope net strung along the side of the ship and into small launches. I had to clamber my way down with one hand, using the other to clutch the rope around my box of white mice. During the voyage I had kept the box beneath my bunk, adding reams of toilet paper each day as bedding for the mice. This was, therefore, the first time the mice had come into public view, and I could see the astonished faces of the marine troops lining the rail overhead, calling to each other and pointing at the box. All I remember, apart from the general sounds of wonderment, was one comment: "Now I've seen every fuckin thing!"

We went ashore and settled down partway up the slope of a hillside, and immediately found that we needed foxholes, not because of enemy fire, but to protect ourselves against the antiaircraft fire from our own ships, aimed at the occasional Japanese planes diving in just over our heads. Our first discovery was that Okinawa's earth was made of sweet potatoes. Everywhere we dug was cultivated, also generously fertilized with night soil—a rich source, we later discovered, of typhoid and paratyphoid bacilli, which a month later produced a mild outbreak of fever among our troops (at first misdiagnosed as scrub typhus because of red spots over the abdomens of some of the patients).

There was, as it turned out, no scrub typhus in Okinawa, nor had there ever been. Nor was there any record of schistosomiasis on the island. This meant that Shope and I were effectively out of business from the outset; I would have to find other uses for the mice, and Shope, after a week or so of walking through island streams looking for the snail vectors without success, had to look for other interests.

Weeks later, in June, the real problem arrived, predicted by no one, a considerable shock to everyone in Island Command Headquarters. An outbreak of a malignant form of encephalitis occurred in several Okinawan villages at the northern end of the island, and three cases of a similar brain disease were reported in American soldiers. In its symptoms and clinical course, the illness was obviously some sort of viral infection of the central nervous system. The likeliest candidate was the agent known as Japanese B virus, which had long been recognized as an endemic disease on the Japanese mainland but was not anticipated by American epidemiologists as a hazard on Okinawa. Our unit set to work on it in a hurry. We obtained specimens of brain tissue from autopsies of several fatal cases, froze them in dry ice, and sent them back to the Guam laboratories by marine air transport. Simultaneously, we inoculated some of our dwindling stock of white mice with brain tissue suspensions. Within a week we had the virus, reproducing the disease in mice, and the Guam serological laboratory reported that it was indeed Japanese B. The cases were isolated, mosquito control measures were taken by Island Command, and, although sporadic cases were seen each week during the rest of the summer, there were no new patients with encephalitis among the American troops. Our job then became the search for the source of the virus. Similar encephalitis viruses were familiar in America, called "equine" encephalitis, carried by horses and wild birds and transmitted to man by mosquitoes. We traveled in jeeps from one end of the island to the other, collecting blood specimens from various domestic animals and wild birds (we were staffed by ornithologists as well as mammalogists and entomologists, and equipped with shotguns), trapping mosquitoes, and driving twice each day to the marine airstrip to load our specimens for quick shipment back to Guam.

Apart from the research, the main concern was security. We set up our laboratory in an abandoned teahouse at the edge of the village of Nago, lived on C rations, and placed bunks in a nearby barn. The day after we had everything in place, we were visited by the local marine commandant and informed that we were on the perimeter of defense; indeed we were the perimeter. Just to the south of our encampment was a narrow road leading between two high hills, and for about a week we received each evening a scatter of machine-gun fire from isolated groups of Japanese soldiers who had made their way through the main line of battle in the southern part of the island. Nobody was hit, but we dug very deep foxholes outside our tents.

The marines assigned a machine-gun post in the road just behind us. We felt the deepest affection for the gunners and, one evening, an even deeper respect. A trip wire had been placed across the road, a couple of hundred yards to the south, and late one evening its flares went off, revealing a group of people running toward us in silence. The young marine posted at the machine gun held his fire and began calling on his radio to the other posts in the vicinity, "Hold your fire. Civilians." Then he climbed over his sandbag barrier and ran down the road in the dark, alone, unarmed, to escort the group—several old men and some women carrying children—up to our barn. I don't suppose anyone recorded this act, but if we had had the authority we would have voted him a gold medal. As it was, we clapped our hands and cheered and told him he was a great marine.

When the war ended I was was living in a tent at the northern end of Okinawa, finishing up the work on Japanese B encephalitis. We had gotten enough evidence from the serological studies of horses on the island to be reasonably sure that these animals were a potential source of the virus, presumably transmitted from horse to man by mosquitoes, but we had seen no sick horses anywhere, nor were there any local stories of a disease resembling encephalitis among the island's animals. Still, almost all the adult horses on Okinawa had high levels of complement-fixing antibodies against the Japanese B virus in their blood samples, while young colts had low levels or no antibody, signifying a virus infection of epidemic scale in these creatures.

We succeeded in persuading Island Command to construct a tightly screened barn with a small laboratory space at one end, and three young colts, very scrawny but otherwise healthy, were brought in from a nearby pasture. We injected each with a saline suspension of infected mouse brain, then took daily specimens of blood from the jugular vein and assayed them for virus by injecting various dilutions of the blood into the brains of mice, using four mice for each dilution. For three days the results were negative, but in the samples drawn on the fourth and fifth days we had the answer we were looking for: although each of the horses remained in perfect health, as far as we could judge, they were now producing high titers of live virus in their circulating blood, ready for any mosquito that might come along.

With three horses in a barn, there were all sorts of studies still to be made: Where was the virus coming from within those horses? What was happening to the brain tissue? Were there lesions similar to those in the human disease? In the middle of planning, our orders came through to wind things up and come back to Guam. We couldn't leave the colts, since each of them contained enough free virus to start up a new epidemic if released. Obviously, they had to be killed. "Sacrificed" is the conventional jargon; experimental animals are almost never killed in biomedical reports, they are sacrificed. We had no idea how to go about killing three young horses. We asked advice from the veterinary officer attached to the military government unit, and he suggested shooting. No, I said, we'll have virus-loaded blood all over the place. "Have you got some formaldehyde?" he asked. We had lots. "Inject some intravenously," he advised. It worked instantly, like magic. Island Command then arranged for a deep grave, and the bodies were safely interred, embalmed with enough formaldehyde for perpetuity, and we packed our bags for the flight back to Guam on the next marine transport that had two empty seats, early the following morning. Fifty miles off Okinawa, one of the engines stopped functioning, and we drifted back to the airstrip and waited there another day for it to be fixed, then off again and straight to Guam.

This was in late September; the war was officially ended, and we expected to go home immediately. Every morning we lined up in front of the bulletin board for news of our departure date; a few people were booked to leave, but the rest of us were told that it would take some time, maybe a few weeks, and advised to settle down. Actually, it was three months. There were several dozen rabbits still left in the animal house, and no experiments planned for them, so I went to work. I had several strains of group A hemolytic streptococcus in the freezer in my laboratory and decided to immunize some rabbits with a mixed vaccine consisting of heat-killed streptococci and a homogenate of normal rabbit heart tissue, to see whether the microorganisms could somehow stimulate the animals to produce antibodies against their own hearts. It had been proven fifteen years earlier by Alvin Coburn that a group A streptococcus infection was essential for the initiation of acute rheumatic fever and rheumatic heart disease, and many people had speculated that the streptococci might have the special property of inducing autoimmune reactions of various sorts. With time on my hands, and an abundance of rabbits, I set it up.

It worked marvelously well. Indeed, I'd never done an experiment before, nor have I done one since, with such spectacular and unequivocal results. All the rabbits receiving the mixture of streptococci and heart tissue became ill and died within two weeks, and the histologic sections of their hearts showed the most violent and diffuse myocarditis I'd ever seen, with clusters of inflammatory cells closely similar to the characteristic lesions of rheumatic myocarditis in humans. The control rabbits injected with streptococci alone or with heart tissue alone remained healthy and showed no cardiac lesions.

I was entirely confident that I had solved the whole problem of rheumatic fever. By this time it was mid-December, and my orders to go home turned up just as I'd finished the last autopsy on the last rabbit.

I can make this a very short story. When I got back to the Rockefeller Institute, in January 1946, I showed my slides to Homer Swift, who was then in charge of the institute's research programs on rheumatic fever and a considerable eminence in the field. Dr. Swift was exceedingly enthusiastic and encouraging, letting me know that I had, in his opinion, duplicated the benchmark pathology of rheumatic fever in my rabbits. I went to work immediately, using the same strain of streptococci as in the Guam experiment, and using the Rockefeller stock of rabbits (the same line that we had brought to Guam in 1944). For the next nine months I set it up, one experiment after another, over and over again, using several hundred rabbits, varying the doses of streptococci and heart tissue in every way possible, and I never saw a single sick rabbit, not an instance of myocarditis, not even the slightest degree of inflammation in any animal's heart.

After a while I gave up trying. The only explanation for the triumphant Guam experiment that I can think of is that there must have been a latent virus of some sort in the Guam rabbit colony, with the property of causing myocarditis, but why it should have done so only in the presence of heat-killed streptococci and heart-tissue suspension is more than I can make up a story about.

I still have my notebooks with those experiments, unexplainable and unpublishable, reminding me that things tend always to go wrong in research. If I had been less confident and more cautious at the time, before leaving Guam I would have taken pains to bring home some of the Guam rabbits, or at least some samples of frozen heart tissue, but I didn't; I was so sure that the experiments meant what I wanted them to mean that it never once occurred to me that I might not be able to get the same results in New York. I had all the controls I needed; I wasn't bright enough to realize that Guam itself might be a control.

# 10
# *Itinerary*

In the 1950s, when the National Institutes of Health were being rapidly expanded in Bethesda, and their research programs were exploding into medical schools all across the country, many of these schools began to build new facilities and add new faculty members to accommodate the boom. My generation, the equivalent of postdoctoral fellows just back from the war, found many more opportunities for research posts in new, well-equipped laboratories than had ever been available before the war, and we began to travel. It was a time for academic tramps: salaries were still at the marginal prewar level (a biochemist friend at Rockefeller used to complain bitterly that the man who drove the Drake's Cakes truck made twice his pay), but the budgets for research equipment, technicians, and experimental animals had never been so generous. We moved around a lot.

Beryl and I moved more than most people in those years,

and acquired moving as a habit. Even when we felt established within a city, roots beginning to go down and all, the addiction drove us to pack up and move to quarters on the other side of town, over and over. Every now and then I have had to fill out long forms for security clearance by the FBI when appointed to consultant positions for the government; these forms have a space of several lines requesting the appointee to write in the addresses and dates of all previous residences since graduation. I cannot do it. At one time or another we have lived in seven different apartments in New York, two houses in Baltimore, two in New Orleans, two in Minneapolis, and three in New Haven, losing chairs, tables, lamps, and rugs at each move. The only safe things were the books and bookcases; we added more shelves and volumes in every town, and Beryl has never forgotten the name of a single book or its place on a shelf. We've never had to clean the books; the library has dusted itself every year or so.

Of all the cities, Baltimore is the place Beryl and I remember with the most affection, although we stayed only two years. The first year we had a flat in an old brownstone on Park Street, across town from the Hopkins medical laboratories. The upstairs tenant was Elliott Coleman, professor of writing, speech, and drama at the university and, not on the side but as the central work of his life, a poet. We became close friends, dining at the faculty club on the Homewood campus and getting to know, through Coleman, more of the Hopkins humanities faculty than we knew of the medical school's. Evenings, Elliott would come down to our quarters to read his poetry or to ask questions about science; my experimental rabbits later turned up, obscurely, in several of his sonnets.

In the second Baltimore year we moved to a much-coveted

small house on Washington Street, a classical Baltimore row house with a white front stoop, just behind Johns Hopkins Hospital. It was a half-block walk to the laboratory and a great convenience for late-night trips to record the results of experiments that couldn't be timed for an eight-hour day.

I might have stayed at Hopkins forever, I suppose, but my appointment was in pediatrics, which I liked but didn't know much about except for infectious disease, and I hankered to get back to internal medicine or neurology. In 1948, Tulane University announced the installation of a new research division of microbiology and immunology in the Department of Medicine and invited me to come and run it. They agreed that I could work on allergic encephalomyelitis, an experimental-animal disease resembling multiple sclerosis, if I would manage the infectious disease clinic in Charity Hospital at the same time. We moved to New Orleans in June, first into one floor of a huge house in the Garden District, later into an apartment in a former army barracks on the university campus. Air conditioning had not yet reached New Orleans except for window fans in the houses and, here and there, small but real air-conditioning units in the laboratories at the medical school, where temperature control was necessary for research. We liked the people, and parts of the city (but not all, not Bourbon Street or Mardi Gras), and the medical school was moving into the first rank of institutions in the South. Beryl and I talked of staying, settling down, even went off with real-estate agents to look at bungalows, but we missed the winter and snow.

Two years later we settled that need, once and for all. Irvine MacQuarrie, the head of pediatrics at the University of Minnesota, came to New Orleans for a visit, carrying an offer of a Chair of Pediatrics and Medicine, really a sort of settee, with laboratories in the newly constructed Heart Hospital attached to the medical school. I could work on whatever I liked—endotoxin, encephalitis, the Shwartzman reaction, streptococci, whatever—provided I would also organize a research unit for both the pediatrics and medicine departments in the new hospital. Beryl and I paid a couple of visits to Minneapolis in the late spring and early summer, admired the city and its climate, were dumbstruck by the new research laboratories then being installed, and agreed to come. We drove north that autumn along the Mississippi in an English Austin, just big enough for us in the front and three daughters and a Welsh terrier in the back, and arrived in Minneapolis in time for an early winter.

The transition from Louisiana to Minnesota was exhilarating for the climate change, but also for other reasons. Minneapolis was, in 1950, as I think it still is, an immaculately clean, bright, and energetic city, prosperous and ambitious to establish itself as the metropolitan center for culture in general in the Northwest. The museums, the Minneapolis Symphony under Antal Dorati, the just-then-organized theater, and the magnificent university were all objects of local pride. It felt like a good place from the day we drove in and settled for a few weeks in a university residence, and Beryl and I went house hunting. Within a month we found, and bought, a wonderfully insulated and storm-windowed clapboard house in St. Anthony Park, on the border between Minneapolis and Saint Paul. It was the first house we had ever owned, the first time each of the children had her own room, the first time I'd had to shovel snow. Roots at last.

We were there from 1950 until 1954, liking everything, even the snow. The laboratory drew in some of the best young bacteriologists and immunologists from around the country: Robert Good, Chandler Stetson, Richard Smith, Floyd Denny, Lewis Wannamaker, Richard von Korff, all interested in rheumatic fever one way or another and all with research projects of their own.

It was the greatest fun, and I thought we would stay there forever.

# 11
# NYU PATHOLOGY

I had been at Minnesota four years when the phone rang. It was Colin MacLeod, professor and chairman of the microbiology department at New York University College of Medicine. He was an old friend of old friends, having been at the Rockefeller Institute in the years just before my arrival. I had the greatest admiration for him as a scientific statesman, and also as a working scientist. He was a member of the Avery-MacLeod-McCarty triumvirate, whose work had opened up DNA genetics. The pathology department at NYU needed a new chairman, he said. Von Glahn had just retired, my name was on the search committee's list, he knew I wasn't a card-carrying pathologist, but most of my work had been in experimental pathology up to then so I ought to know something of the field, would I be interested? I said yes, I'd come in a minute. He said, well, the search committee, you know, has a list to go through. I said, yes, I know, but if they get down to my name I'll come in a minute.

The NYU College of Medicine was located in a series of battered, late-nineteenth-century buildings along First Avenue just north of Twenty-sixth Street, across from Bellevue Hospital. Its reputation was based in part on great Bellevue, whose wards were the sole source of clinical teaching material for NYU. It was also known to be a school largely and traditionally populated by students from New York City itself, many of them from relatively poor families, mostly Jewish, some first-generation Italians, a few Irish Catholics, a very few blacks—a different student body from those at Columbia and Cornell. The school had turned out in the past some spectacularly famous people—James Shannon, Jonas Salk, Albert Sabin, Joseph Goldberger—but its solidest reputation was for its production each year of intelligent, soundly trained, above all *Bellevue*-trained, physicians who formed the backbone of medical practice in New York City and its immediate environs.

I knew some of the faculty personally, and more of them by reputation. This accounted for the alacrity of my response to MacLeod's call and my eagerness to accept the job before it was even offered. Homer Smith in physiology, Severo Ochoa in biochemistry, Bernard Davis in pharmacology, Donal Sheehan in anatomy, MacLeod in microbiology. I knew of no medical school in the country with a basic science faculty to match the roster at NYU. In the clinical departments they had William Tillett in medicine, John Mulholland in surgery, Samuel Wortis in psychiatry and neurology, William Studdiford in obstetrics, Marion Sulzberger in dermatology, and Emmett Holt in pediatrics. Currier McEwen, a rheumatic disease specialist, was the dean. It was a small medical school, living in extremely cramped quarters, dependent on an underfinanced and overburdened Bellevue for its clinical teaching, drawing students only from its local community, and backed by a huge parent university known to be in chronic financial difficulties. And, at the same time, it seemed to me the best faculty and the most interesting and exciting medical school in the country. A new building was being constructed for laboratories and teaching areas across the street, just north of Bellevue, and there were plans for a new University Hospital.

A few weeks later I went to New York to be interviewed by the search committee, which consisted of most of the department chairmen from the NYU medical school. There was a conversation of a couple of hours. I was asked what I would do with the Department of Pathology if I were selected to head it. I knew the department was very small, having been depleted by recent retirements and resignations, and I replied that I would try to build it up, if I could, with research people skilled in experimental pathology (which included immunology, by my lights), and I would hope that the department's teaching would include a fair amount of exposure of medical students to the major areas of ignorance in medicine. I said I thought the pathology department, poised as it was midway between the basic sciences and the clinical disciplines, was in a good position to pay special attention to ignorance along with its other more obvious responsibilities.

Several weeks later, in the spring, I was told by Currier McEwen that the pathology chairmanship was mine if I wanted it, and my wife and I settled down to make plans for moving to New York that summer. Up to then I had not discussed the budget with any of the NYU people. It was limited, I knew, and would need a substantial supplementation from NIH grant sources if the kind of expansion I had in mind was to be possible. My own salary, I learned in sorrow, would be somewhat less than I was getting at Minnesota, and we knew that living expenses in New York were going to be considerably higher. Still, the opportunity seemed even more fascinating than when I'd first learned of it, and we began exploring possibilities for housing in New York.

Our next-door neighbors in Minnesota, the Nafes, had moved to New York a few months earlier. John Nafe, professor of geophysics, had taken a position at Columbia and was working at the Lamont Observatory, fifteen miles north of the city on the west bank of the Hudson River. We got in touch with Sally Nafe, and she began house hunting. Just north of Lamont, built into a cleft in the Palisades, is the village of Sneden's Landing, an antique Dutch settlement made up of around thirty-five houses, built more or less at random on either side of a winding road leading from the 9-W highway down to the river's edge. One of these, an old but recently remodeled Dutch farmhouse, was coming up for rent shortly, and Mrs. Nafe found out about it. Sneden's Landing had been resettled long since by a mix of artists, writers, theater people, and a few scientists, and we were extremely lucky to find a vacant house; lots of people from the city wanted to live there, and Beryl and I made up our minds on the instant of hearing about it long-distance. We packed up the house while I finished the experiments still in progress in the laboratory, and moved in the summer of 1954.

Sneden's was wonderful, nothing less. The elementary school was a short walk up the hill and across the highway. Alongside the school was the general store and the Palisades village library, to which generations of good readers had contributed good books for all ages. Our daughters, now aged thirteen, ten, and six, enrolled with enthusiasm and started school in September. The school activities were the center of interest for the village, and most of the teachers were local residents. Everyone went to the baseball games between Palisades and the other elementary schools in Rockland County. Our oldest daughter, Abigail, developed into a fair pitcher, a regular member of the team. I remember one game held on a Wednesday afternoon when the Palisades team was losing badly because of the county regulation that religious study was an option for all classes on Wednesday afternoons, and the team had been depleted of its best players. Toward the end of the game, losing gamely but embarrassingly, the team suddenly shouted in elation, "Here come the Catholics," and so they did, running across the field flushed with virtue from their catechisms and ready to take on the world, which they then did, winning the game handsomely. Ever since, "Here come the Catholics" has remained alive in our family's language, used whenever we felt ourselves disadvantaged but with hope just ahead—an unexpected check in the mail, for instance, was greeted with "Here come the Catholics."

The only thing wrong with Sneden's was the distance between the upper Palisades on the Hudson and East Thirtieth Street on the East River. There was no easy way. For a few weeks I indulged in the fantasy of becoming a regular, methodical commuter: Beryl would drive me to the railroad station in Tappan a few miles away, I would board the 7-something train, with briefcase and newspapers, ride the long slow trip to Weehawken, take the ferry across the Hudson, then a bus across Thirty-fourth Street and to the laboratory, late all the time. I gave it up and took to the car, south to the George Washington Bridge and down through Manhattan, an hour or more each day. Finally I joined a car pool and spent the next four years averaging four hours a day in gossip, which I calculated came to two solid months of the year sitting in an automobile.

The first responsibility of a new department chairman is recruiting, which usually involves the much more difficult (sometimes impossible) job of getting rid of people already on the scene. The latter was not a problem: the Department of Pathology at NYU had already become very small due to attrition during the preceding few years, and the people who had stayed were, perhaps by a process of natural selection, among the best in the business: John Hall, a master of surgical pathology; Marvin Kuschner (now the dean at Suny–Stony Brook); and Robert McCluskey (now chief of pathology at the Massachusetts General Hospital). These men, together with a very small group of junior associates, were the core of the department. A year later, Sigmund Wilens, who had left NYU because of illness a while back, returned to take on the directorship of the pathology service in Bellevue.

The department's quarters in the new Medical Science Building at Thirtieth Street and First Avenue were adequate in square footage but designed as though exclusively for the use of people looking down microscopes at slides. All of the rooms were narrow cubicles, with two benches for microscopes and a sink. The laboratory equipment on hand for the department's research consisted, by actual inventory, of one water bath, one incubator, several dozen copper racks for test tubes, and two venerable centrifuges. Luckily, the walls were easy to knock out, since the wiring and pipes ran through the ceilings, and the benches were movable, so that within a few months we had some wide laboratories that could be used for more general research purposes.

I had expected recruitment to be very difficult. Pathology was at that time not a particularly promising discipline for basic researchers, even those engaged in experimental pathology with animal models, and the limited space available, the relatively low salaries paid by NYU, and the difficulties involved in family living in New York City or its suburbs all made it seem hard to attract the people we needed.

I had not counted on the same magnetic force that had brought me to NYU by a single telephone call. MacLeod, Smith, Ochoa, A. M. Pappenheimer, Davis, and the rest were attraction enough, and I began to get letters from some of the very people I had hoped to induce to come, asking me what I was up to and what the place was like. The first to come was Chandler Stetson, who had worked with me as a postdoctoral fellow at Hopkins, had gone on to the Rockefeller, and then into the army to do research on streptococci in the rheumatic fever laboratories at Fort Warren, Wyoming. I asked Stetson on the phone if he'd like to consider coming, and his answer was that he'd be there in a month.

Next I received a long-distance call from Leo Szilard, the nuclear physicist, from his laboratory at the University of Chicago. He had heard that I was planning a new kind of pathology department, he said, and I should get hold of Howard Green, who was doing interesting immunological work with chick embryos at the Walter Reed Hospital in Washington. I called Green, and he agreed to come sight unseen.

Later in the year I went to an immunology conference in Paris and met Baruj Benacerraf, a young immunologist working at the Broussais Hospital. Benacerraf had been raised in Paris, but had come to New York during the war, gone to college at Columbia, to the Medical College of Virginia for his M.D., to Kabat's laboratory at Columbia for postdoctoral work in immunology, and then back to Paris. We had a long lunch together, and by the time coffee arrived he had agreed to join the department as assistant professor, at a salary of $5,000 per year.

These people became attractions for others, and within the next two years we had assembled a remarkably bright group of young investigators, most of them working on immunologic aspects of human disease mechanisms. At the same time, visiting investigators on sabbatical leave began coming to the department to work: Philip Gell from the University of Birmingham, Guy Voisin from Paris, Jaroslav Sterzl from Prague, Jacobus Potter from Edinburgh, and Peter Miescher from Basel. Miescher, a specialist in experimental hematology and immunology, came back for several visits and finally decided to stay on at NYU in a permanent position (he went back to Switzerland in 1966 to head up the hematology clinic at Geneva).

It was an exciting period for pathology, a field with a long history of meticulous study of morbid anatomy, preoccupied for many years with detailed analysis of the morphologic changes which characterized disease. At just about the time when we were organizing the new department at NYU, it was being recognized that the mechanisms of disease not only were the proper responsibility of academic pathology departments but were now becoming approachable by scientific experimentation; this was particularly evident in what had previously been the quite separate discipline of immunology. The presence of a concentrated population of young investigators in this field in a single department soon became of interest to the NYU medical students, and we began receiving a steady flow of students in our laboratories, working after hours, weekends, and through vacation time as voluntary assistants. This was probably the most important change of all for the department. Nothing so stimulates research as the presence of students, and the NYU students were the most interesting and most insatiably curious any of us had seen. One result was the production of a new generation of NYU graduates committed to research in immunology, many of whom achieved outstanding careers later on: Emil Gottschlich, who later devised vaccines against meningococci and gonococci at Rockefeller University; Gregory Siskind (now a professor of medicine at Cornell); Frederick Becker (now director of cancer research at the M. D. Anderson Center in Houston); Ronald Herberman (a research chief at NIH); and a long string of others first did research as medical students in the Department of Pathology.

The laboratories in the Department of Pathology were soon filled to overflowing; they were small rooms to begin with and outfitted for less than half the people who finally signed on, but it is my impression that the work went better for the crowding. No investigator of whatever seniority had more than a single room, about 400 square feet, at his or her disposal. There were occasional agonized complaints: people kept bumping into each other; there were traffic jams at the instruments, especially the pieces of high-cost, shared equipment; visitors had no place to sit down; and the offices, including mine, were alcoves measuring 6 by 4 feet, but the research kept coming. Everyone took part in the teaching, lecturing in the morning to the entire second-year class or walking from bench to bench in the teaching laboratories through the afternoon, but there were enough faculty people on hand so that the teaching responsibilities were not burdensome for anyone. The teaching laboratories and the autopsy rooms in the ancient Bellevue Hospital morgue were the two centers of teaching activity, always filled with students.

At first, the interns and residents who came to the department for postdoctoral training were young men, and a few women, who were headed for careers as hospital pathologists. These professionals were in short supply, and a community hospital pathologist (who usually had the added responsibility of running all the diagnostic laboratories) was assured of a useful career at a salary near the top of all medical specialties. Later, a couple of years after the department had been reorganized, we began receiving applications from candidates who planned to stay in academic medicine and wanted training as professional pathologists and at the same time as researchers. Still later, an even larger number of people came after finishing internships and residencies elsewhere in medicine, in order to learn immunologic research.

At the outset, there were some complaints about us within the national community of professional hospital pathologists, and a few hostile articles and editorials in the pathology journal. It was felt by some that we were skewing or trying to skew classic pathology away from its traditional objectives, training young people to do research instead of practicing pathology, downgrading the field in general. My response to these criticisms was that the problem of human disease mechanisms had been on the agenda of pathology since the time of Virchow in the nineteenth century, and we were not departing, as claimed, from that tradition. It was true, however, that we were not attempting to produce new generations of hospital pathologists; we would not have been very good at this if we'd wanted to, nor would more than a few of our students, and there were plenty of other medical schools training such people.

In the end, nevertheless, the record of experimental pathology as a source of teachers and investigators in universities was an encouraging one. In the eight years I spent in the immunology laboratories at Minnesota and NYU, ten of the younger people who passed through in the course of their training went on later to become the chairmen of departments at major universities, including Harvard, Yale, Florida, Texas, Mississippi, Minnesota, Northwestern, and New York University.

A year before the end of my term in pathology, the NYU medical school received a major grant of $750,000 from the Commonwealth Fund, with instructions to use the money in whatever way seemed best for the future of the school. The preliminary discussions with the foundation, and within the faculty, had concerned the possibility of installing some new educational programs for paramedical personnel—physiotherapists, occupational therapists, medical social workers, and other professionals who work in close collaboration with physicians. We did a lot of arguing, and a rump session of the chairmen, led by Colin MacLeod, began meeting to discuss alternatives. We emerged several months later with a different plan, for which I had the job of drafting the initial proposal, a new program to be called the Honors Program, under which all of the new money would be set aside for the single purpose of providing new opportunities for selected medical students to undertake fundamental research while still in their student years, with fellowship support, ample time off from the regular curriculum, and the requirement of a formal, full-scale thesis before graduation "with honors." The

Honors Program won the day, after much debate, and was initiated with its base of operation and offices in the Department of Pathology. It was the first program of its kind that I am aware of, and later received support from NIH as the precursor of similar programs set up at other universities, culminating in the combined M.D.-Ph.D. training programs now under way in about twenty medical schools. Looking back, I think this was the most interesting experiment of all the ones I had a hand in during my time at NYU.

# 12
# NYU Bellevue Medicine

In 1958, after four years as chairman of the pathology department, I was shifted into a different chair and a different world. Dr. William S. Tillett, who had been the chairman of medicine and director of the NYU medical division in Bellevue, retired that year at age sixty-five, and a faculty search committee was appointed to find a successor. I was not made a committee member, which puzzled me some, because the research linkages and interdependence between the departments of medicine and pathology had always been close, both by tradition and by the deep interest of both groups in human disease mechanisms. I thought my department ought to have representation in the decision, and I was about to approach the committee to protest the matter when the committee approached me instead and asked if I would like to take the job.

I objected, but very mildly, for fear they might be convinced by any arguments I might make. I pointed out that I was by that time a long way away from clinical medicine, which was obvious, and that my previous experience had been as much in pediatrics and neurology as in medicine. I said the obligatory things about how the committee ought to look further afield to find someone better qualified, but then I said, loudly enough, that if they really wanted me to do it I'd move in a minute.

Not that I was discontent in pathology. I wasn't; I loved the job and the department and was having the time of my life, and I could happily have stayed a pathology professor all my life. But the chance to run the clinical services in Bellevue Hospital was simply beyond resisting.

Bellevue was rather like the Boston City Hospital, an ancient set of buildings beginning to fall apart, long open wards filled with the sickest and poorest of New York's citizens, inadequately supported by the city but obliged, unlike any other hospital in Manhattan, to take in all patients who came to its doors. In 1958, the medical services were divided among NYU, Columbia, and Cornell; NYU had responsibility for the third and fourth divisions, occupying four floors of wards in the old C&D Building on the south side of the Bellevue campus facing the East River. Columbia had the first division and Cornell the second, each with two wards. There was a lively, sometimes bitter rivalry among the three schools for space and prestige, but the other two directors, Dickinson Richards of Columbia and Thomas Almy of Cornell, were universally respected figures in academic medicine and also good friends of mine. The only hard problem was the other NYU medical service, the fourth division, which had always been managed by the postgraduate medical school, a separate part of New York University with a clinical staff based some blocks away in the old Postgraduate Hospital. It had been used for many years to teach refresher courses to practicing physicians who came to the wards for periods of several weeks throughout the academic year, but it had never had medical students roaming its wards, nor did it possess much in the way of space or facilities for research. My job was to organize a single NYU medical service, to combine the fourth division with the third, and to use all four floors—about 120 beds—to teach medical students and house staff. It seemed a risky business at the outset, feelings waiting to be hurt at every turn. There were two separate missions involved: undergraduate medical education and postgraduate teaching, seemingly incompatible with each other; also, the two professional staffs were potentially at odds with each other, one devoted to fundamental aspects of scientific medicine with ambitious plans for building more extensive research laboratories, the other committed almost entirely to the practice of medicine. Moreover, there were problems ahead of the very worst kind, having to do with the economics of medicine: plans were already being drawn for a brand-new hospital for private patients to be built just to the north of Bellevue, into which the old Postgraduate Hospital would be moved with its name changed to University Hospital. Would this be for the private patients of the physicians on the fourth division, or the third, or both, and how would the beds in the new hospital, on which doctors' incomes would surely depend, be apportioned?

There was another problem. One of NYU's most difficult tasks, not shared by Cornell or Columbia, was the management of another set of wards within the Bellevue Psychiatric Center, called the psychomedical division. These were the beds occupied by patients who were simultaneously psychotic and medically ill—schizophrenics, manic depressives, alcoholics, and senile patients who were brought to Bellevue with pneumonia, meningitis, heart disease, diabetes, or any other condition requiring the skills of internists as well as psychiatrists. These wards had to be integrated somehow with the other parts of the NYU medical service and used for teaching and research.

It seemed an enormous job, and neither I nor any of my friends in the NYU faculty was sure it could be done. It was assumed that there would be academic squabbles all over the place and resistance on all sides to any sort of change. As it turned out, it was remarkably easy, and I held the position for eight years, a record for me at that time. I only left it, and then with reluctance, to become dean of the medical school.

What made it work was the quality of the people involved. Although engaged in quite different commitments and obligations, the senior members of the two staffs were, with a few exceptions, extraordinarily skilled in their lines. Moreover—and this made all the difference—most of them had been trained and brought up in Bellevue, and had chosen to stay for their careers in that hospital. My closest friends—and the best clinician-scientists I have ever encountered—were Saul Farber, a noted kidney specialist, and Sherwood Lawrence, the discoverer of the "transfer factor," an important component in immune reactions. These men, and their colleagues, were in love with Bellevue—the whole hospital. The two departments became a single working unit, almost overnight.

It may also have been an advantage for me to have come into the chairmanship from such a mixed background in academic medicine. I was not conceivably a threat to anyone. It was conceded that I knew something about infectious disease and immunology, but unimaginable that I could turn myself overnight into a Geheimrat ready to dominate all branches and specialties within NYU medicine. Also, it was clear enough that I was not likely to convert myself magically into a master practitioner and take over the beds in the new hospital.

Gradually, over the next few years, the department expanded with some coherence into a fairly large organization with clinical and research disciplines of solid strength. There were several floors of unoccupied rooms, some of them bedrooms used long ago for patients with tuberculosis, and these were converted into serviceable research laboratories. Other laboratories were made available in the medical school building itself and, when finally constructed, within University Hospital. It became a big department for the time, with a house staff of more than sixty interns and residents and a somewhat larger number of postdoctoral fellows and research assistants. By 1966, the department's group of young clinical investigators and teachers were turning up as selected speakers on the podium of the Young Turks (the American Society for Clinical Investigation) at the annual May meetings in Atlantic City, in numbers to match their competitors in the other research-oriented universities elsewhere in the country. To have a paper on the program in Atlantic City was the most important event in an academic career in medicine in those years, a firm step up life's ladder. It still is, although when that city became the center for gambling casinos, the meetings were scheduled in other cities; it is still officially announced, each year, that the "Atlantic City meetings" will be held in San Francisco, or Washington, or wherever.

During my time in Bellevue, the ward rounds in which I took the greatest interest were on the psychomedical service. These wards had previously been regarded by the teachers, house staff, and students as a Siberia, but they became, on close scrutiny, a gold mine for teaching problems, easily the most intellectually stimulating area in all the vast reaches of Bellevue Hospital. Patients who were brought in with what seemed to be commonplace, everyday illnesses, neglected because of mental incapacitation in their homes or on the New York streets, turned out to have one exotic disease after another superimposed on mental illness or even, on occasion, problems like thyroid disorders hidden away as the cause of the mental illness. I have never seen, before or since, so many different manifestations of chronic alcoholism, most of them brain syndromes due to alcohol itself—delirium tremens, Korsakoff's syndrome, Wernicke's disease, polyneuritis, acute alcoholic hallucinations, aphasia. I had been taught, long ago, that most of these were irreversible and that nothing could be done to change the course of events once alcohol had started dissolving away the brain. It was a constant surprise to see some of these patients gradually emerge from their version of hell, get well, and, clean and neatly dressed, leave the hospital, most of them on their way back to the Bowery despite the strident warnings and finger-pointing exhortations of the house staff and nurses. I knew one middle-aged man who was in and out of the psychomedical ward ten times over a three-year period, each time with profound aphasia and total loss of memory; each time, with nothing but bed rest, nourishment, and good nursing care, he was transformed back into an intelligent, thoughtful but rueful, good-humored man, on his way once more to the skid rows of town.

Bellevue served as the city's hospital for sick prisoners sent in from Rikers Island or the House of Detention in Manhattan. Some of these people were known to be dangerous, and beds were provided for them on locked and guarded floors in the psychiatric building, known as the psychoprison wards; others, convicted of minor crimes, were admitted to the open wards of the NYU services, always accompanied by police officers, who were stationed at the bedside in rotation throughout the day and night. The police seemed to like this leisurely duty, but the interns were always indignant at what seemed an infringement of their authority and a waste of city money. I remember one such patient on a ward I rounded, an elderly, obese man caught pickpocketing in Times Square who had gone into congestive heart failure while serving time on Rikers Island. I examined him, propped up inside his oxygen tent and breathing what looked plainly like his last, while an armed cop stood sharply on duty looking over my shoulder. It seemed to me that both of us were performing a futile ritual at a death bed. My contribution surely could not help, his certainly wasn't needed to keep the poor man in bed. Luckily, digitalis and diuretics were unexpectedly effective, and after several weeks the man was out of his tent, able to sit in the chair alongside his bed, chatting amiably with his police guards. Finally he was discharged and taken back to jail to finish his short sentence, and that was the last we saw of him until six months later, when he arrived back on the ward, once again in an oxygen tent, arrested again in Times Square for pickpocketing. We lost track of him after that admission. For all I know, he may still be at it.

For an intern, Bellevue was a unique place for learning how to provide the best of all possible medical care for the sickest of patients under the worst of conditions. If you could make it at Bellevue, you could thereafter cope with almost anything. One of my interns, a quiet, reserved, very bright graduate with top marks, vanished from the hospital a week after beginning service; the police found him unconscious in a midtown hotel room where he had gone to kill himself with Nembutal. He was revived, and we took him back, after he explained apologetically that he had been afraid of not being good enough, afraid of making mistakes. Reassured vehemently, he went back to work and later turned out to be one of the best doctors in the hospital.

Every kind of medical equipment, every item from penicillin to toilet paper, was in short supply and often lacking altogether. Interns on one ward learned to swipe from other wards in emergencies, also to hoard supplies. Everything needed repairing: beds, wheelchairs, litters, radiators, windows, elevators. Especially elevators, malfunctioning or out of commission almost half the time. When Dickinson Richards retired from the Columbia division, we gave him a dinner at the Century Club, and after the speeches we presented him with a bronze paperweight, cast to duplicate the elevator buttons in the A & B Building and their accompanying notice: one button stamped UP, the other DOWN, and a small sign reading PUSH UP FOR DOWN.

The trouble with Bellevue was not so much money as the way the money was doled out. There was a central administrative office in Bellevue, but New York City's Department of Hospitals acted as actual day-to-day administrator, as it did for each of the municipal hospitals, from offices several miles away at 125 Worth Steet. Every request even for minor expenditures, surgical gowns for instance, had to be threaded through anonymous bureaucrats in the department offices, and most of these required initialed approvals by other officials in the city's Department of Purchase and the Bureau of the Budget in different city office buildings. Actually, the total budget assigned to Bellevue at the beginning of each fiscal year was ample for the hospital's needs, but it could only be spent line by line from downtown. The real budget was always much less than the amount allocated, in some years 25 percent less. Bellevue served as a sort of piggy bank for the city: $35 million would be appropriated on paper, but with the expectation of spending $25 million in that year, leaving the city $10 million to draw on for snow removal or subway maintenance or potholes.

It was an awful system, and Richards, Almy, and I spent many hours down at Worth Street arguing bitterly with the commissioner of hospitals, whose regular response was that he was on our side but could not change anything. On several occasions we ended up in the mayor's office, explaining carefully that our patients' lives were jeopardized by the way Bellevue was managed. One day I met a high-perched city official at a committee function. I thought I had finally reached the center of the administrative demonology and launched our appeal for help, but he displayed equanimity: Bellevue had all the money it needed, he said, and added that the city should not be spending any more on a place "filled with bums." I spluttered and said something in a raised voice about bums being a small minority of Bellevue's patients, the hospital was filled to overflowing with men, women, and children who were there because they were poor, and then I made the case for changing the system to give the hospital enough authority to make its own decisions about spending, even on a smaller scale than the official budget. I got nowhere. The budget bureau, I was coolly informed, had been at this for many years, and how long had I been a professor?

I became, overnight, an outraged, offended reformer. I wrote indignant letters, made speeches, caught the ear of any elected official I could find, even went for a secret meeting with the Democratic Party boss of the day, Carmine De Sapio, to complain about the System (someone had told me this would be a good idea; it wasn't).

One day I learned the name of the functionary in the budget bureau whose initials always appeared on Bellevue's canceled budget lines, and I went down to the Municipal Building to see who he was and what he was like. His desk was one of two dozen in a huge room filled with clattering typewriters and steadily ringing telephones. Stacks of paper covered the desk. He was a hugely fat man with the kindest of faces and a courtly, avuncular manner, leaning back in a sturdy swivel chair with a red pencil (red!) in his hand. What could he do for me, he asked. I explained that I was a Bellevue director of medicine, concerned that the hospital was broke, worse than broke. Ah, he said, that was something he understood, and he was working hard to fix it. We had a brief, unsatisfactory conversation. He explained that he had been at that desk for twenty-five years, having come into the position from a career as a civil engineer, and he knew all about hospitals. Good luck to you, Doctor, he said, as I walked away from his desk.

We decided to organize. A friend of Almy's, a professional public-relations man, volunteered his services and put together a new, tax-exempt society named "The Better Bellevue Association," dedicated to two principal objectives: to persuade the city to get on with a long-deferred plan to rebuild the entire hospital and, at the same time, to wrench Bellevue out of the hands of the downtown bureaucracy and provide the institution with some sort of autonomy. The first was the easy part. Architectural plans had already been drawn up, and all that was needed was to turn up as a delegation before the Board of Estimate and declaim in favor of haste. At that time, in the early 1960s, such a huge public-works project had many political attractions regardless of cost, and anyway, no one possessed an accurate notion of what the cost might be. The official estimate was $50 million; when the elegant new hospital was finally finished, twelve years later (nine years behind schedule), the cost was probably three times the original guess.

But the second mission was a lot more difficult, even, as it finally turned out, impossible. At the end, the City Charter itself got in the way. There was simply no legal mechanism for freeing Bellevue from the fingers and red pencils of the overseeing agencies in lower Manhattan; everybody would be in jail if we tried. At last, we developed a new plan and carried it north to Albany. It was a design for a piece of legislation to create a quasi-public corporation, patterned after the Port Authority of New York and New Jersey, which would own and operate the municipal hospitals but also mandate that each institution would be, administratively, a tub on its own bottom. The central feature of the bill was the lifting of the whole system clean out of City Hall, with all the funds to come from Medicare and Medicaid. At the time, in the late 1960s, it seemed entirely feasible. The bill sailed through the state legislature, but at the very last minute, just before it was ready for passing, someone inserted the provision that the mayor would appoint the president and board members of the new corporation, thus effectively retaining the same control over the hospitals as before. What then happened was the creation of a new, highly centralized bureaucracy, duplicating the former Department of Hospitals, running the hospitals from the same distance. It was a discouraging enterprise. I still think it might have worked as a public authority. Even with the cutbacks in health-care funding since the 1970s, it would work better today than the present system.

When I became dean of the NYU medical school in 1966, I moved out of the hospital and up the street, but not completely out of Bellevue's business. There was still a severe housing problem, affecting both Bellevue and the school. All the adjoining blocks were slum properties, with many abandoned, decaying buildings but no places for married interns, nurses, or technicians to find living quarters. The Phipps Foundation took up the problem, and back we all went to City Hall with a proposal to clear the seven-block area between Twenty-third and Thirtieth streets for the construction of new apartment buildings. By the late 1960s the plans had been accepted, and within a few years Bellevue South emerged, meeting the needs of the local residents as well as the Bellevue community.

Today, Bellevue is a spectacular new building, a huge white square dominating the East River south of Thirtieth Street. Columbia and Cornell have left, and NYU is now responsible for the whole place, with excellent arrangements for sharing certain clinical and technical responsibilities with University Hospital up the block. I regard it still, as I did when I first walked through the unhinged doors of the old building, as the most distinguished hospital in the country, with the most devoted professional staff. If I were to be taken sick in a taxicab with something serious, or struck down on a New York street, I would want to be taken there.

When I drive past, or think of it at a distance, I have two sharp memories. One is of an early-morning session with the interns in 1959, when I had just begun as professor of medicine. A young intern was presenting his report on a patient who had been on one of the wards for two weeks with advanced pneumonia and meningitis. He had been up all through the previous night, doing everything he could think of and enlisting help from the senior physicians and consultants in infectious disease, but the patient had died. Halfway through his formal presentation tears appeared in his eyes and rolled down his cheeks, and he wept while he finished. I knew that these were tears not of frustration but of grief, and I realized, for the first time, what kind of hospital I was in.

My second memory is of Mrs. Marjorie Barlow, a New York great lady in her eighties who had charge of the patients' library. Each day, Mrs. Barlow, my wife, and several other women, all volunteers, set forth on the Bellevue wards with book carts for the patients. Mrs. Barlow, a tall, willowy, fragile-looking lady, a Horace scholar, also an authority on Hroswitha, and knowledgeable about everything ever written on boxing, reserved one set of wards for her own book cart, never allowing anyone else to take it on. It was the psychiatric prison ward, where she always succeeded in making new friends. She had a high regard for the patients on those wards and never seemed frightened, although there were plenty of frightening patients. For some of them, she had words of affectionate respect: "He is a good reader, you know," she would say.

# 13
# THE BOARD OF HEALTH

Before I went to medical school my knowledge of the public health profession was limited to the childhood singsong: "Marguerite, go wash your feet, the Board of Health is down the street." I learned very little about public heath in medical school, beyond what was called "The Sanitary Survey," a required field study in the summer between the third and fourth years, in which all students were assigned in pairs to a municipal or county health department as observers. I had a classmate friend who lived in Cincinnati, and we arranged to survey that city. It took two weeks to do and another week to write a report. We learned more than we had known about sewage disposal, water bacteriology, venereal disease clinics, premarital Wassermann tests, and public baths, but it seemed a long way from medicine, a longer way from science. That was the extent of my training until I was appointed in 1956 as a member of the Board of Health of the City of New York, on which I served for fifteen years.

The New York Board of Health is the oldest such body in the country, having been organized in 1866 to deal with the epidemics of cholera and yellow fever then plaguing the city. It was statutorily set up as a separate legislative institution, empowered to write its own laws relating directly to matters of public health in the city; these laws comprise what used to be called the Sanitary Code, now known as the Health Code. The board is made up of five members—the commissioner of health, three physicians, and one layman.

I was appointed to the board to succeed Dr. Thomas Rivers of the Rockefeller Institute, who had just retired. The other members were Dr. Haven Emerson, representing public health, Dr. Samuel Z. Levine, then professor of pediatrics at the Cornell medical school, and Chester Barnard, head of the American Telephone and Telegraph Company. Later on, Louis Loeb, a lawyer in Manhattan, Dr. Walsh McDermott, professor of public health at Cornell, and Paul Hays, professor of law at Columbia, became board members. For most of the fifteen years of my service on the board, Dr. Leona Baumgartner was the health commissioner.

At about the time I became a member, the Board of Health, and the city's health department itself, began to run out of things to do. Long before, in the last quarter of the nineteenth century and the early years of the twentieth, the New York City Department of Health had set the style and pace for other American city and state health departments, and its record of accomplishment and innovation was long and distinguished, but now, in 1956, the major problems of public health which had engaged the energies of the department for so many years seemed to have been solved. Tuberculosis had become a relatively uncommon disease, and the formerly elaborate measures for case finding and tracking people in contact with active cases were no longer needed. Syphilis and gonorrhea were still important matters, but now they had become more like therapeutic than preventive problems, or so it seemed. The great epidemics of poliomyelitis, typhoid, scarlet fever, and diphtheria, formerly requiring the quarantining of schools and homes all around the city, had come under control. Smallpox had turned up briefly a few years earlier among returning travelers, requiring the most massive vaccination program ever launched in any city, but there had been no epidemics of this disease, or plague, or cholera, or anything else. The whole Department of Health, whose traditional mission had always been primarily to maintain constant vigilance against contagions in the populace, had to examine its usefulness for the future.

There were a few traditional functions which the Board of Health would just as soon have set aside, but these had become embedded in the economic life of the city. Milk, for example. In earlier decades the department had been obliged to keep a close eye on the milk industry because of the risks of tuberculosis and streptococcal infection in cattle. This had meant an intricate system of inspection at all levels—at the farms, in the depots and bottling companies, on the trucks, and in all the dairy shops of the city. The dating of milk containers had been introduced as a device for limiting the amount of bacterial growth in milk, not just as a way of protecting the consumer against sour milk. All of these things cost a lot of money, not only for the inspection and sampling work by the department, but also for the added personnel needed by the producers and distributors of milk. In view of the fact that milk no longer seemed a significant vector for disease in New York, the commissioner and Board of Health proposed backing out of the milk business and issued a public statement to that effect. This was followed by an uproar from all sides—the milk producers, the unions of milk handlers and truck drivers, the shop owners, and any number of civic organizations, all demanding that public hearings be held. Lawyers sent in long briefs and petitions protesting the idea of changing anything concerning milk.

The Board of Health met on the first Wednesday morning of each month, and for nearly all meetings during the first year of my appointment the agenda was crowded with milk items. I remember one morning when the dating of milk bottles was under discussion and a lawyer representing the delivery-truck drivers was addressing the board. His real concern was that if milk dating was modified or given up, the drivers would have fewer jobs to do because of fewer trips to pick up outdated containers, but he didn't put his case that way. Instead, he strode up to the platform, shook his fist at us, and shouted, "Gentlemen, can you actually prove that stale milk is not a cause of cancer?" We never did get much done. Each change in the law, even minor ones, resulted in new public hearings and appeals to the courts, and the milk laws stayed pretty much as always.

Fluoridation of the city's water supply was an even harder and more emotional problem for the department. The evidence was indisputable by the mid-1950s: adding trace amounts of fluoride to drinking water had been proven to furnish solid protection against dental caries, and was also known to be entirely harmless. It seemed the most reasonable of all public health measures. New York had already stalled for too long a time, and the city's need was greater than most cities'. The majority of children in the poorer sections of town had never been seen by a dentist. Tooth decay and the premature loss of teeth were serious health problems for hundreds of thousands, with nothing being done about them. The department viewed the prospect of fluoridation with zeal, for here was an opportunity to be professionally useful. The Board of Health passed the *pro forma* resolution mandating the installation of fluoridation, the mayor concurred, and then the roof began falling in. Public hearings were demanded by one after another civic group and their lawyers, protesting what was called an attempt to poison the public, and it looked, for a while, as if the matter would be tied up in the courts for years. In its early stages and throughout the debates, the issue took on an ideological aspect which became increasingly bitter. At one long hearing, starting in midafternoon and running through midnight, held in the auditorium before the Board of Estimate, the members of the Board of Health were accused by impassioned speaker after speaker of being Communists or tools of Communism. In many parts of America, fluoridation was seen as a foreign plot to weaken the country, possibly by setting loose an epidemic of cancer. Fluoride was un-American. This may have been the issue that ultimately turned the tide: in came the medical and dental societies, the nursing organizations, representatives of the New York Bar, and finally the business lobbies, who were persuaded that the city as a whole could save a lot of money by achieving sound teeth for its citizens. Late one night, after the last of the hearings, the mayor and the Board of Estimate approved the appropriation of a considerable sum of money for the necessary equipment, and the Board of Health held a quiet celebration for itself.

Other problems were identified by the health department

staff as new areas for public health reform, and lists were drawn up for consideration by the board. All the conventional concerns were on the lists—restaurant and market sanitation, meat inspection, heroin addiction, cigarette smoking, alcoholism, rats and cockroaches, periodic physical examinations for the citizenry, free clinics for childhood immunization, and so forth, but at the top of the agenda—the one thing the department staff would really have wanted to do something about if it could get at it—was housing. Jerome Trichter, a long-time professional in the department and a genuinely devoted public servant, arranged several tours for the Board of Health to take a direct look at the kinds of quarters people lived in, in Harlem, the South Bronx, Bedford Stuyvesant, and Brownsville. Even then, back in the late 1950s, it was obvious that massive portions of the city's housing stock—especially the "old-law" tenements, in which most of the city's poor resided—were standing ruins. It was also obvious that they were a menace to the health of all tenants. We traveled on several winter days from block to ravaged block in a long gleaming black city limousine, feeling like guilty intruders, climbing in and out of tenements and old multistory apartments, guided by Trichter, who had carefully arranged the tour so that we could see the worst problems at first hand, along with department photographers taking pictures so that we would not forget what we were looking at: lightless staircases with broken treads to cause long falls in the dark, toilets flooded and leaking continually into apartments below, broken windows in corridors, broken boilers in the basements, rats as big as cats, roaches as big as rats, and every kitchen jam-packed with small children trying to keep warm around a lighted stove, no heat in any other room. The stoves, burners going day and night, carrying obvious hazards of fire and carbon monoxide poisoning, were the only technology for making do. The stoves, and also a variety of ingenious contraptions, principally iron poles, for locking and blocking the front doors of every apartment against thieves looking for money for heroin. This was nearly twenty-five years ago. So far as I know, the only improvement since then has been the final destruction and leveling of many of these buildings in the South Bronx and Brownsville, and probably the crowding of their tenants into other parts of the city, probably now just as bad. It was made clear then to the health department by other parts of the 1950s bureaucracy that this was not a real health problem, not the business of the department, and was being looked after by other city agencies. Anyway, even back then the money was beginning to run out.

The Board of Health had its attention drawn to other matters, more readily approachable and conveniently trivial. An outbreak of hepatitis in Brooklyn was traced by epidemiologists to a tattoo parlor on the waterfront catering to incoming sailors, and this was turned, for a few weeks, into a major undertaking requiring the writing of new laws. There were more debates, public hearings, and legal documents. Could regulations be written to assure the sterilization of the tattoo needles, or should the establishments be closed down once and for all? This was an easy one. At the end, no convincing arguments could be put together for any sort of public benefit from tattooing, and even though the risk of hepatitis was not entirely convincing, the tattoo parlors were outlawed. It was a small, extremely small, satisfaction, almost no satisfaction at all.

There exists in the law one great power assigned to the Board of Health as its very own, not shared with any other agency of the city or state. This is the power to declare a State of Imminent Peril. Originally, in the nineteenth century, this was intended to legalize the taking of extraordinary steps by city officials to limit the spread of epidemic disease, quarantining whole areas, stopping the movement of citizens, even locking people up if necessary. Before the government could do things like this, the Board of Health had to declare that, in its professional opinion, imminent peril existed for everyone.

Twice during my term, the board was under strenuous pressure to make the proclamation, both times to prevent highly inconvenient but hardly perilous strikes by labor unions. The first was a strike by gravediggers for better pay. During the long winter of this strike, coffins were stacked in great columns in every city cemetery. The Board of Health was asked by the other authorities to declare that this would endanger the public safety, especially when the spring thaws came. We were unable to see why, although it was plainly a cause of great dismay and personal anguish for many families. Luckily, the strike was settled soon enough to keep the matter away from public debate or open hearings. The second time the issue arose was in the course of a very long strike by sanitation workers. Garbage piled up in plastic bags and rubbish cans all over the city, covering the sidewalks and spilling out into the streets. Everyone in town was inconvenienced at first, then repelled, finally infuriated by the mess and smell. If there was ever an Imminent Peril, we were loudly told by the press, this is it; please declare it! We thought it would be better (and fairer) to have the strike settled in some other way, rather than set so risky a precedent, and said that we knew of no epidemic disease likely to be started by garbage. We were, at least until the strike was finally settled, the most unpopular five people in city government.

The research arm of the New York Department of Health was the Bureau of Laboratories, organized in the late nineteenth century by Dr. William H. Park, then commissioner, in order to do work on such problems as diagnostic tests for diphtheria, tuberculosis, and typhoid, and also to study the pathogenesis of contagious disease in general. During Fiorello La Guardia's administration in the 1930s, the Public Health Research Institute was set up as an independent institution, with funds from the city treasury, to undertake long-term basic research on the human diseases regarded as major public health problems, particularly tuberculosis and nutritional diseases. In 1959, Leona Baumgartner decided that the research activities of the health department needed reappraisal and expansion, and also some clearer view of what would be needed for the long-term future. She organized a two-day meeting of about a hundred biomedical-science and public health authorities from around the country, with the help of Drs. Frank Horsfall, Walsh McDermott, James Shannon, then director of the National Institutes of Health, and me. The outcome was something of a surprise, an agreeable one for the deans and professors of the medical schools in New York City. Instead of advising, as had been expected, an expansion of the Public Health Research Institute or the Bureau of Laboratories, which the conferees regarded as near-perfect for their legislated functions at their present size, the conference recommended that the health department organize a new mechanism for sponsoring scientific research in the medical institutions of the city. Mayor Wagner accepted the idea, and announced that $8 million a year, a dollar for each citizen, would henceforth be invested in medical research. The Health Research Council was the outcome.

This council became an extremely useful social invention, the envy of universities and their medical schools in other parts of the country. Its principal function was to select and financially support young investigators who wanted to join the faculties of the New York medical schools, with five-year fellowships which could be renewed for a second term. The effect on the schools themselves was almost instantaneous: applications came in from as far off as California from young men and women who, as it turned out, had always hoped to live and work in New York. It was a surprise. The city's institutions had never had funds enough to engage in much recruiting beyond their walls, and the city, for all its great size and seven respected medical schools, ranked a shade behind Boston, Baltimore, and Los Angeles as a center for biomedical science. Within a few years, Mayor Wagner's dollar-per-citizen began to make all the difference. We used to calculate, for the purpose of encouraging continuation of the Health Research Council at each year's budget hearing, that each dollar had at least a tenfold multiplying effect. The new HRC fellows were highly successful in the competition for grant support from the NIH, and each grant provided jobs for technicians, laboratory assistants, and other workers. The Health Research Council actually created a new and booming industry for New York. I believe that this single institution moved New York City into unarguable first place among the nation's scientific medical centers.

Although the city's officers never insisted that the Health Research Council restrict its sponsorship to research projects directly relevant to the city's health problems, and a large amount of valuable basic research was accomplished, unrelated to any special disease, everyone involved was aware of the general intention of the fellowship fund. Deliberate efforts were made, therefore, to launch work on several practical problems. One of these was heroin addiction.

By the early 1960s, heroin was generally agreed to be the public health problem of greatest concern to the city, and a formidable economic problem as well. More than five hundred deaths occurred each year among adolescents and young adults alone, and a considerably larger number from homicides and accidents related to heroin. Acute hepatitis spread through the addict population and then beyond, due to contaminated needles. Bacterial endocarditis, pulmonary disease, malaria, and chronic liver disease were attributed to heroin use. Babies were born addicted, at high risk of early death. As to the cost, it was figured that well over $1 billion was lost to city residents each year from thefts to raise funds for heroin. Each addict required at least $50 a day to sustain an average sort of habit, at the prices of that time.

I had become interested in the problem and written some worried essays about it, because of seeing so much of it at first hand among adolescents in the Bellevue wards. I was appointed chairman of an ad hoc committee of the Health Research Council charged with studying the matter and returning several months later with recommendations. But this was in the spring of 1962, at a time when I had just accepted a sabbatical appointment as visiting professor of bacteriology at the University of Edinburgh, starting in July. The committee met several times, during the spring, finding the heroin problem more complicated at each meeting and finding, as well, very few openings for real research work. We were not even sure it was a medical problem as much as a social dilemma. I persuaded Eric Simon, a young Ph.D. biochemist in the Department of Medicine at NYU, to take an interest in opiate addiction as a biochemical problem, and placed a laboratory and fellowship at his disposal. It was a considerable gamble for him; to leave his present line of work and take on a brand-new, obviously intractable problem was a high risk. He took it on with some zest, however, and kept at it for the next twenty years. Simon spotted, early on, the best of openings: was there a special receptor for morphine and its derivatives somewhere in the brain? Using compounds labeled with radioactive tracers injected into rats, he discovered that specific receptor cells do indeed exist, in particular cells of the mid-brain, and that the attachment of the drug to the surfaces of these cells could be prevented by morphine antagonists. The work of Simon and a few others led, eventually, to the discovery that the brain manufactures its own morphine-like compound, endorphin, to fit with those same receptors.

My other, still more indirect contribution to the heroin problem was to go off to Edinburgh for my sabbatical on schedule. This meant that someone else would have to serve as chairman of the ad hoc committee in the year of my absence. I telephoned Vincent Dole, a friend of mine and an elegant investigator at the Rockefeller Institute with broad interests and experience in metabolic disease, and asked him please to become chairman. He demurred at first, pointing out sensibly that his work was nowhere near the heroin problem and he knew nothing about the matter. But then he agreed to chair the meetings, and off I went to Scotland. During that year, Dole served first as a conscientious chairman, then as a fascinated student of the field, and, before the year was out, as an irresistible, unstoppable scientist at work.

He dropped his other research and plunged into laboratory investigations of all aspects of heroin addiction, emerging before long with the idea—and then a clinic for testing the idea at Rockefeller—that methadone might be the ideal drug to block the episodic highs of heroin addicts without other side reactions, thus enabling the addicts to get off heroin and on to other ways of living. Dole's work still stands as one of the stunning successes of clinical research, and remains the most practical and effective way of dealing with heroin addiction.

As a sort of spin-off, Dole and his wife, Marie Nyswander, became interested in the city's prisons. I visited the House of Detention in downtown Manhattan with Vincent one afternoon several years later, where he was running a methadone clinic, and discovered that he had good friends on every cell block. Prison life had turned into a separate, half-intellectual, half-emotional concern for Dole. I suspect that sooner or later he will turn his attention more closely, and emerge again with a good idea for changing the jails. It would fit nicely into the original objective of the Health Research Council—to enlist good scientific minds for worrying about matters crucial for New York's future—but in the meantime the council languished and died for want of money at the time of the great New York budget crisis in the mid-1970s. A ghostly form of the institution floated off to Albany and still exists, on paper, as a state council, but the city has lost it, perhaps for good. I hope not, though. When the money returns, if it does, and the city once again becomes affluent enough to be ambitious, and if anyone asks me, I would vote to restore the Health Research Council—still at a dollar per citizen—before doing anything else.

# 14
# Endotoxin

Dr. Oliver Wendell Holmes once laid out the dictum that the key to longevity was to have a chronic incurable disease and take good care of it. Even now, 150 years later, this works. If you have chronic arthritis you are likely to take a certain amount of aspirin most days of your life, and this may reduce your chances of dropping dead from coronary thrombosis. When you are chronically ill, you are also, I suppose, less likely to drive an automobile, or climb ladders, or fall down the cellar stairs carrying books needing storage, or smoke too much, or drink a lot.

In research, something like the Holmes rule holds if you readjust the words. The key to a long, contented life in the laboratory is to have a chronic insoluble problem and keep working at it. But this does not mean staying out of trouble. On the contrary, it means endless, chancy experiments, one after another, done in puzzlement. It is worth it, for this is the way new things are uncovered, whether or not—and usually it turns out not—they illuminate parts of your problem. But nothing ever gets settled once and for all when you work this way.

The best example I know of is the near-century-long, still uncompleted study of endotoxin, in laboratories all around the world. I know about it at first hand, for I have profited from it in my own laboratory because of having trapped myself into starting work on endotoxin thirty-five years ago and having never succeeded in stopping, never even settling conclusively any major part of the matter at hand.

Endotoxin had its beginning as a biomedical problem with the first attempts, early in this century, to make a vaccine against typhoid fever. Typhoid vaccine quickly became famous, but not for any remarkable capacity to prevent typhoid fever. Indeed, although it remains in more or less routine use to this day, in much the original form of a crude suspension of heat-killed typhoid bacilli whose walls are rich in endotoxin, its effectiveness has always been marginal at best. The most spectacular and deeply interesting property of typhoid vaccine has, from the beginning, been its capacity to cause fever. The term "pyrogen" was instantly coined, and the massive literature on what started out as an inconvenient side issue began to accumulate. Today, the bibliography of scientific papers dealing with endotoxin and its biological and chemical properties numbers in the tens of thousands. From time to time long reviews of the field are written, usually from one or another specialized point of view—for example, the chemical structure of the lipopolysaccharide molecule in the bacterial cell wall, which is, in fact, the endotoxin—and the accumulated reviews themselves comprise a formidable body of reading, far beyond any single reader's endurance.

For all the intensity and scale of the work, it is still not known how endotoxin causes fever. Even more puzzling is the array of other intoxicating effects, ranging from scattered hemorrhages throughout many organs of the body to overwhelming shock (similar in its manifestations to traumatic shock) and death. Many of the studies have been done with rabbits, which are the most susceptible of conventional laboratory animals, but it is known from observations in human beings injected with tiny quantities of endotoxin, and also from the reactions of patients suffering from typhoid or related infections, that man is even more vulnerable, perhaps the most sensitive to endotoxin of all animals. Less than one millionth of a milligram will produce shaking chills and high fever in a normal adult.

The center of the puzzle is that endotoxin is really not much of a toxin, at least in the ordinary sense of being a direct poisoner of living cells. Instead, it seems to be a sort of signal, a piece of misleading news. When injected into the bloodstream, it conveys propaganda, announcing that typhoid bacilli in great numbers (or other related bacteria) are on the scene, and a number of defense mechanisms are automatically switched on, all at once. When the dose of endotoxin is sufficiently high, these defense mechanisms, acting in concert or in sequence, launch a stereotyped set of physiological responses, including fever, malaise, hemorrhage, collapse, shock, coma, and death. It is something like an explosion in a munitions factory.

This is the reason, or one reason anyway, why the problem of endotoxin is so engrossing. It provides a working model for one of the great subversive ideas in medicine: that disease can result from the normal functioning of the body's own mechanisms for protecting itself, when these are turned on simultaneously and too exuberantly, with tissue suicide at the end.

You can observe this sort of thing in one of the tricks that experimental pathologists have played with endotoxin for many years, known as the Shwartzman phenomenon since its original discovery by Gregory Shwartzman in 1928. A small quantity of endotoxin is injected into the abdominal skin of a rabbit, not enough to make the animal sick, but just enough to cause mild, localized inflammation at the infected site, a pink area the size of a quarter. If nothing else is done the inflammation subsides and vanishes after a day. But if you wait about eighteen hours after the skin injection, and then inject a small non-toxic dose of endotoxin into one of the rabbit's ear veins, something fantastic happens: within the next two hours, small, pinpoint areas of bleeding appear in the prepared skin, and these enlarge and coalesce until the whole area, the size of a silver dollar, is converted into a solid mass of deep-blue hemorrhage and necrosis.

The Shwartzman reaction is a kind of pathologic ritual, with tight rules. The skin must be injected first, then the intravenous injection, and the time interval between the injections is crucially fixed at between eighteen and twenty-four hours. When the intravenous injection is made earlier or later, nothing happens. If the injections are made in reverse, first by vein and then in the skin, nothing happens. If both injections are made into the skin, nothing happens.

I heard about this phenomenon when I was a third-year medical student at Harvard in 1936, and began reading everything I could find about it and about endotoxin. I began to think up things I'd like to do with it, and became obsessed with possibilities. Here was a way that a particular area of tissue could undergo its own devastating disease at its own hands, so to speak, by its own devices. It required only a special combination of events, each of which, by itself, was harmless. It seemed a model for all sorts of events that might be occurring in the infectious diseases then commonplace and insoluble problems for medicine. I wondered, among other things, what would happen to a rabbit if both injections were given by vein, spaced eighteen to twenty-four hours apart. Could the internal organs, any of them, be "prepared" for the reaction? And then, of course, I made the usual illuminating discovery, the commonest epiphany in research: someone else had done my experiment. Two German pathologists, several years earlier, had given both injections by vein, and the outcome was an extraordinary disease called bilateral cortical necrosis of the kidneys. The photographs were impressive; I can still remember them on the right-hand page of the pathology journal: the outer portion of both kidneys entirely destroyed by dense black zones of hemorrhage and necrosis. This phenomenon is now known as the *generalized* Shwartzman reaction.

Several days after seeing that paper, I had the experience which turned my mind toward research in experimental pathology. It was a Thursday afternoon, and I was sitting at a conference table in Professor Tracy Mallory's office at the Massachusetts General Hospital, at the weekly seminar in Mallory's elective course in advanced pathology. I forget what was being talked about, but I remember leaning back in my chair, bumping my head against a heavy glass jar on the shelf of tissue specimens behind me, and knocking it over. I picked it up to replace it, and saw that it contained a pair of human kidneys with precisely the same lesion as the one in the photograph. The label said that the organs were from a woman who had died in eclampsia, with severe bacterial infection as well.

The M.D. program was not then, and still is not, very satisfactory training for research in biomedical science. Then, as now, the Ph.D. program provided a much more rigorous and profound experience in science, with a better grounding in the basic fields of biology needed for medical research. Earning an M.D. has, however, one enormous advantage which makes up in part for its deficiencies. After four years of medical school it is impossible to think about a problem in biology, or to read a paper, without having part of one's mind trying at the same time to make connections with human disease. The intrinsically amazing aspects of the generalized Shwartzman phenomenon were enough by themselves to catch my interest and make me hanker to work on it, but the human tissue in that glass jar sent me over the line, and from that day on I was resolved to turn myself, one way or another, into an investigator of that queer reaction.

Death is the most familiar of events in living organisms, long before the creature as a whole dies. If you were a conscious member of the replicating mass of developing cells in an embryo, you would probably be horrified by the wholesale slaughter of cells on every side. The piecing together of a fetus involves a great deal of obsessive editing. Structures are meticulously put together, like the forerunner of the kidney, and then as though in an afterthought these cells are destroyed and a new, more advanced kind of kidney is installed. In the process of building the modular columns of nerve cells for the cortex of the brain, many more neurones are produced than can be fitted into the needed circuits, and these redundant cells must be killed off before the final, perfect electronic device can be put in operation. During all of adult life, dying and replacing goes on at a great pace in many vital organs; the lining cells of the intestinal tract, the blood cells, and the skin are the busiest, but there is even one class of proper brain cells, the olfactory neurones in the lining of the nose, where death and replacement by new cells takes place at about three-week intervals.

With such mechanisms at work everywhere, it would be no surprise to learn of occasions when the self-destructive devices are fumbled with, turned loose at the wrong times, with disease as the outcome.

In the Shwartzman phenomenon, cell death is caused by a shutting off of the blood supply to the target tissue. After the second injection, the small veins and capillaries in the prepared skin area become plugged by dense masses of blood platelets and white cells, all stuck to each other and to the lining of the vessels; behind these clumped cells the blood clots, and the tissue dies of a sort of strangulation. Then the blood vessels suddenly dilate, the plugs move away into the larger veins just ahead, the walls of the necrotic capillaries burst, and the tissue is filled up, engorged by the hemorrhage.

The same sort of sequence takes place in the kidney in the generalized Shwartzman phenomenon. The smallest vessels, those in the glomerular capillary tufts, are plugged by tiny thrombi, the surrounding cells die of oxygen deprivation, the vessels themselves burst, and bilateral cortical necrosis is the end result.

My colleagues and I tried our hand at sorting out the participants in these catastrophic events, over the next ten years, at Johns Hopkins, Tulane, and the University of Minnesota. The first nice observation was that when the circulating white blood cells are lifted out of the picture, just before the expected occlusion of vessels, the phenomenon is completely prevented; this is accomplished by treating the rabbits with precisely the dose of nitrogen mustard needed to eliminate the leucocytes for a period of twenty-four hours. Next, we found that the temporary inhibition of blood clotting, by a properly timed injection of heparin, prevented the phenomenon. Cortisone didn't work, which was an odd thing in view of the fact that cortisone will totally protect animals against the lethal, shock-producing effects of endotoxin.

We never did succeed in learning how the white blood cells contribute to the phenomenon, beyond their obvious role in blocking the flow of blood, nor did we find out how the blood vessels were induced to burst their walls.

But in the course of trying, we ran into a number of other things, unrelated, as it turned out, to the Shwartzman phenomenon but interesting enough in themselves, worth, as the Michelin travel guide says, a detour. One of these was the surprising action of papain. It occurred to us that the release of a proteolytic enzyme by damaged tissue cells might be one way of rupturing small vessels, and we guessed that such an enzyme might be the sort most active in the reduced acid environment which we knew existed in the rabbit's prepared skin. So, without much hard thinking, we injected small amounts of a plant enzyme of this type, papain (from papaya latex), into rabbit skin, and within an hour had a fair copy of the hemorrhagic necrosis of the local Shwartzman phenomenon.

This, we thought, was surely the way to go. The next step was to inject papain intravenously, in order to reproduce the generalized reaction, kidney necrosis and all. We did this, and nothing happened. The animals remained in good shape, active and hungry, and their kidneys were unblemished. We repeated it, using various doses of papain, always with negative results. But then we noted that the rabbits, for all their display of good health, *looked* different and funny. Their ears, instead of standing upright at either side, rabbit-style, gradually softened and within a few hours collapsed altogether, finally hanging down like the ears of spaniels. A day later, they were up again.

I am embarrassed to say how long it took to figure out what had happened. I first observed the papain effect in 1947, and examined microscopic sections of the affected ears without finding anything wrong in the cells, fibrous tissue, cartilage, or other structures in the ear, and dropped the matter. Every few months I returned to it, sometimes in order to demonstrate the extraordinary change to friends and colleagues, but never with any sort of explanation. It was not until six years later that it dawned on me that since a rabbit's ears are held upright by cartilage there simply had to be something wrong with the cartilage plates in those ears. I went back to it, comparing the quantity of cartilage matrix—the solid material between the cartilage cells—in the ears of papain-treated rabbits with normal rabbits, and found the trouble immediately: although the cartilage cells themselves appeared perfectly healthy, almost all of the supporting matrix had vanished after papain. Moreover, the same changes occurred in all cartilaginous tissues, including the trachea, bronchial tubes, and even the intervertebral discs. Parenthetically, several years after my published report on this business, some orthopedic surgeons introduced the use of papain as a method of getting rid of ruptured intervertebral discs without surgery. Beyond this, so far as I know, nothing of any practical value ever emerged from the work on papain, but there were one or two points of theoretical interest, perhaps relating to disease mechanisms.

Papain works as an enzyme only in its reduced state, while the oxidized enzyme is inert. Contrary to our expectations, it was only the oxidized inactive enzyme that worked in rabbits. Intravenous injections of reduced papain caused no ear collapse, nor did the enzyme succeed in getting out of the bloodstream into the cartilage tissue. Only inactive papain could do this, and presumably was then activated by some reducing agent already present in the cartilage matrix. At least two parts of the reaction were under the rabbit's own control: the transport of the enzyme across capillary walls and then its activation.

There is a long scholarly paper by two sociologists, Barber and Fox, which seems to have become something of a classic, reappearing in several collected reviews of the field, about this papain work, but from a specialized sociological point of view. They had learned from a pathologist friend that another researcher had encountered the papain effect on rabbit ears in the course of enzymological studies on another problem and had failed to follow it up. The sociologists' problem was to discover why I had done so and he had not. I remember being interviewed in my laboratory at Bellevue, five years after the work, trying to think of why I had been unable to drop the problem, and all I could think of was that it was so entertaining. By that time I had also learned that cortisone had the remarkable property of keeping the ears collapsed, so I was able to justify working on so seemingly frivolous a problem by the possibility that cortisone might possess the capacity to interfere with the synthesis of mucopolysaccharides in tissues—which gave the whole affair a down-to-earth, usable aspect. But I was obliged to confess, despite this, that the work had been done because it was amusing.

# 15
# CAMBRIDGE

In 1959 and 1960, I took a semisabbatical leave from NYU and Bellevue, three summer months each, and went with my family to England to work on the placenta. Sabbaticals are designed not for resting but for getting into new ground for a while. I was looking for a new kind of problem, different from endotoxin and microbial infection, and I had become enchanted by the strange architecture of the human placenta. Also, my associates and I had discovered a few months earlier that enormous masses of this multinuclear tissue, owned and operated by the developing fetus, were constantly breaking off and floating away into the mother's circulation throughout pregnancy. I was curious to find how this happened and what it meant.

The trophoblast of the human placenta consists of two layers of cells which are the most primitive and at the same time the most specialized of all our cells. They are there from almost the very beginning, soon after the fertilized ovum begins to undergo its successive divisions, forming the bank of invasive tissue which will attach itself to the wall of the uterus and there insert its roots for the future embryo. After implantation, the trophoblast becomes the lining for an immense lake of maternal blood, on which the embryo feeds. From the mother's immunologic point of view this tissue is, of course, foreign, and the trophoblast is, doctrinally, a homograft.

But it is the most successful of all homografts, surviving throughout all nine months of pregnancy. At the outset it is made up of individual cells with single nuclei, called the cytotrophoblast, but soon after implantation these cells give rise to a second layer of fused cells, containing innumerable nuclei, called the syncytial trophoblast. The syncytium expands and extends, becoming one gigantic single cell comprising the bulk of the whole placenta, the largest single cell in nature. Somehow or other, it is not rejected by the mother—not, at any rate, until the end of pregnancy; perhaps then it is rejected all at once, and the act of rejection becomes the act of birth.

It has been known for a long time that fragments of syncytium occasionally became dislodged from the placenta; small bits of multicellular tissue had been found occluding the small veins and capillaries of the lung in women who died in eclampsia, but it was not recognized that this was a normal event. Our group at Bellevue had learned about this by drawing samples of blood from catheters inserted in the femoral vein up to the level where the uterine veins, draining the placental lake, enter the vena cava. Throughout normal pregnancy, from the sixth week on, this blood contained ten or more multinucleated syncytial masses in each cubic millimeter, each one the size of forty or fifty leucocytes. The fragments looked solid enough in our blood smears, but obviously they must have ways of breaking up into smaller bits or disintegrating altogether soon after entering the mother's blood; otherwise they would have blocked the circulation of blood in her lungs and pregnancy would be an impossibility. We made the guess—still a good one, I think—that the dislodgement of this mass of fetal tissue is a way of desensitizing the mother, flooding her circulation with so great a load of foreign antigen that her immunologic system becomes "paralyzed," unable to respond.

I had corresponded about this with Professor J. Dixon Boyd, then head of the Anatomy School at the University of Cambridge, and he liked the general idea enough to invite me over to work in his laboratory. Boyd was at that time one of the leading experts on the morphology of the human placenta, with a large collection of histologic specimens of placental tissue at all stages of pregnancy.

We arrived at Cambridge in May, and found temporary lodgings in a tiny cottage in Grantchester on the grounds of "The Orchard." Later we rented half of the very large Cornford house at Conduit Head, off Madingley Road, just at the edge of town. Thinking back over the years since those two summers, I have trouble recalling all the day-to-day events in the laboratory, but I can remember every single aspect of the Cornford house—the chickens in the side yard, the garden, the feeling of elation each time I drove up the narrow drive, the special Cambridge sky, the quiet.

I began work in Boyd's laboratory, trying to set up facilities for tissue culture in hopes of getting live preparations of trophoblast for immunologic study. Within a few days it became obvious that this was going to be a technical problem beyond my competence, and Dixon Boyd proposed that we talk with Miss Honor Fell (now Dame Honor), head of the Strangeways Laboratories, who knew more about the cultivation of cells and tissues than anyone else in town, or anywhere else for that matter. So, one mid-afternoon I drove out to the Strangeways in a car still half loaded with unpacked bags and went in to see Honor Fell.

She was waiting for me at her laboratory bench, a tall and dignified lady of the old school, very much in command of every detail of the Strangeways research programs, and at the same time totally preoccupied by her own work, which she had always done, and I believe continues to do, with her own hands. She had a capable and alert laboratory assistant, a younger man, whose function was to prepare the materials she was to work with and hand her various items as she worked, but it was clear from the moment I first saw her that Miss Fell did her own research. She may have had an office desk somewhere in her laboratory, but I never saw her at it.

Miss Fell had invented a good many of the intricate procedures for cultivating embryonic organs. The greatest difficulty in this kind of work had been the risk of contaminating the cultures with bacteria from the air or from the investigator's own respiratory tract, and Miss Fell had worked out a meticulous and nearly infallible technique for keeping her cultures free of accidental infection. By the late 1950s it had become easier to prevent infection by using antibiotics in the tissue culture media—a combination of penicillin and streptomycin was then in general use for this purpose. But not in the Fell laboratory. She used no antibiotics at all, and indeed disapproved of them, partly because they introduced new variables into any experiment, but also, I suspect, because she felt that any careful worker should be able to keep things clean and it was somehow unsporting to bypass the need for such care by relying on antibiotics.

We talked about the placenta for about twenty minutes. I had the impression that Miss Fell was not as swept off her feet by the problem as I was, and not at all confident that it would be a suitable topic for a brief summer's work, but she was extremely kind to me and offered some helpful suggestions. Then teatime arrived, and the conversation shifted away from the placenta and, with mounting animation, to Miss Fell's laboratory problems. Chief among these, at the top of her mind, was the cultivation of whole mouse embryonic bones, for which she had worked out a beautiful technique involving the placement of the tiny bones on little doilies submerged in a nutrient medium of great complexity and usefulness. The problem then preoccupying her laboratory was the action of vitamin A on these explants: within a day or so after adding the vitamin, all of the cartilage matrix became dissolved and disappeared, leaving the cartilage cells, still healthy, bunched together in solid clumps. The impression was that vitamin A was somehow inhibiting the production of matrix by these cells, but without doing damage to the cells themselves.

The microscopic slides of the mouse bones were remarkably like those of my rabbits' ears. By good luck, I had brought some of these specimens along from New York, and they were still in an unpacked box in the car parked just outside, so I brought them into the laboratory and we had a close look at the mouse bones and the rabbit ears, side by side. It was as cheerful a tea as I've ever had. By five o'clock we had agreed that there must be a connection of some sort between the systemic action of papain in the rabbit and the local action of vitamin A on mouse embryo bones, and by five-fifteen we had planned the experiment which was to occupy the rest of the summer.

I cabled home the next morning for a supply of highly purified crystalline papain protease, and made arrangements with Professor Boyd for a clandestine allocation of six young rabbits for my use. This was illegal, for I had not obtained a license for animal experimentation in England, but the matter seemed urgent enough for me to risk a penalty; also, I was quite sure that a single experiment with six rabbits would provide all the answer we needed.

During the next few weeks, everything fell neatly into place. Miss Fell added a little papain to her mouse bone cultures, and duplicated the vitamin A effect, while I administered large doses of vitamin A by stomach tube to my rabbits, and their ears collapsed within twenty-four hours. In each case, the microscopic changes in the involved tissue were precisely as predicted.

The almost self-evident explanation was that vitamin A must be causing the release of a proteolytic enzyme in both the mouse bone explants and the cartilage tissues of living rabbits, with the same extremely selective action as papain, removing the matrix without affecting the cells. This would have to mean that a source for such an enzyme, but in its inactive form, already exists in normal cartilage tissue. The best candidates for a source known at that time were the lysosomes, small membrane-lined sacs filled with a variety of hydrolytic enzymes, scattered through the cytoplasm of all cells. These organelles had been discovered a few years earlier by Christian de Duve (who received the Nobel prize for the discovery) and were suspected of playing a role in the digestive processes of individual cells. Over the next few years, the

Strangeways group proved that this was indeed the explanation for the vitamin A effect: the vitamin caused the selective disruption of lysosome membranes and the release of their enzymes, and a protease with properties similar to those of papain then leached away the cartilage matrix, and that was that.

# 16
# THE GOVERNANCE OF A UNIVERSITY

Long ago, in the quiet years before World War II, being chairman of the Department of Medicine (or the Department of Physiology, or Surgery, or whatever) in an American medical school was pretty much like being head of the English department over in the main part of the university. From year to year, the medical school remained the same size, in the same aging buildings. The university shared a fixed portion of its general endowment with the medical school. The latter did its best to secure extra money of its own, but never great sums, from affluent alumni or their patients. The medical school's budget was a fairly steady sum of money from one year to the next. The main difference between running a major department in the two parts of the university was that many of the key faculty members—the professors—in the clinical departments of the medical school were not paid a salary by either the university or the medical school. These men had their own offices elsewhere in the city where they conducted a private practice and earned their livelihood. The holding of a professorial appointment on the medical school faculty was a mark of professional distinction, carrying some assurance that patients would be referred to the physician-teacher's care for diagnosis or treatment by other doctors, and the physician-teacher did quite well.

At the time I reached the low rungs of the academic ladder, ready to try my hand at science, money for research had to be scraped together in very small sums, enough to buy rabbits and mice and minor supplies. Technicians were rare personages, to be found only at the benches of the senior-most professors. It was taken for granted that I would be responsible for preparing my own bacteriological media, washing and sterilizing my own glassware, and looking after whatever animals were involved in my experiments. For the first two years of my work at the Thorndike, the costs of all the supplies came to $500 a year, this from a special endowment called the Wellington Fund, which had been bequeathed to the laboratory years earlier.

After the war, the federal government made the decision that science was useful and important, and research within the medical schools became a much more serious business. As a result, the differences between the management of the medical school departments and their counterparts elsewhere in the university became sharper. While the English department continued to lead its more or less steady existence, assured of a fixed complement of tenure and nontenure positions for its faculty, and equally assured of a steady but very modest fixed income from the university's endowment to pay its costs, the medical school departments, and some of the science departments within the university proper, began to explode. An individual faculty member in, say, the Department of Microbiology, an expert on the streptococcus or meningococcus, could now file an application for a government grant to support his laboratory, purchase supplies and new instruments needed for the pursuit of some new ideas, pay part or all of his own salary, and hire one or two technicians. A few years later, whole departments came under grant support, with money enough to recruit new faculty members, provide fellowship funds for increasing numbers of Ph.D. candidates as well as postdoctoral fellows, even money to build new laboratory quarters. And later still, in the 1950s and throughout most of the sixties, the federal research funds arrived in bundles large enough to renovate most of the buildings in existing medical schools and to build more than a score of brand-new schools.

I served as the dean of one medical school, New York University, during the period of rapid expansion and at another, Yale, during the time when the funds were reduced, not yet to a trickle, but to a thin, even flow. Both experiences brought me closer to the inner workings of the modern university than I had ever been before. Being a professor in a medical school is a good way to find out how these highly specialized institutions and their connected hospitals function, but the university itself seems a remote place, almost nonexistent. Being a dean is different: you get to look inside.

The governance of academic institutions has been considered and reconsidered, reviewed over and over by faculty committee after committee, had more reports written about it than even the curriculum, even tenure. Nothing much ever comes of the labor. How should a university be run? Who is really in charge, holding the power? The proper answer is, of course, nobody. I know of one or two colleges and universities that have actually been tightly administered, managed rather like large businesses, controlled in every detail by a president and his immediately surrounding bureaucrats, but these were not really very good colleges or universities to begin with, and they were managed this way because they were on the verge of running out of money. In normal times, with institutions that are relatively stable in their endowments and incomes, nobody is really in charge.

A university, as has been said so many times that there is risk of losing the meaning, is a community of scholars. When its affairs are going well, when its students are acquiring some comprehension of the culture and its faculty are contributing new knowledge to their special fields, and when visiting scholars are streaming in and out of its gates, it runs itself, rather like a large organism. The function of the administration is solely to see that the funds are adequate for its purposes and not overspent, that the air is right, that the grounds are tidy—and then to stay out of its way.

To function in accordance with its design and intentions, a university must be the most decentralized of all institutions.

This is not easy to do, and it is a surprise that it works, by and large, as well as it does. The risk of politics is always there, and there can be nothing so confusing and aimless as academic politics, or, sometimes, so bitter and recriminatory. The remark was once made, attributed to the California politician Jesse Unruh, that the trouble with university politics, the thing that makes it so different and more potentially damaging than the ordinary politics of state government, is that "the stakes are so low." There is some truth to this. In real life, nothing much is to be gained through the temporary leadership of a committee or *ad hoc* group out to reform the curriculum or change the parking rules or get rid of the president or whatever. The leader, even if successful, is not likely to emerge from victory with a higher salary or more space for his office or laboratory, and he runs a greater risk of not changing anything or, at the most, acquiring the reputation of a disturber of other people's peace.

The worst of all jobs is that of the dean. He carries, on paper, the responsibility for the tranquillity, productivity, and prestige of all the chairmen of the departments within his bailiwick, and when things go wrong at the outskirts, in the laboratories of an individual faculty member or the cubicle of a graduate student, the blame is swiftly transported toward the center, past the desks of the senior faculty, past the chairman's secretary (who is usually the person running the day-to-day affairs of the department), and straight on to the dean.

The actual power of a dean to do anything much is marginal at best and, even at that best, dangerous if he tries to use it. In the major research universities, and especially those with medical schools, the money for running the departments is money generated by the individual faculty members and their students. In good times, this money cascades over the dean's head, arranges itself in rivulets that flow past the chairman's office out into the individual budgets of the faculty. In this circumstance, the faculty members are convinced that the university is run by their efforts, is forever in debt to each of them for its sustenance, and owes them a strict and accurate accounting for every dollar of the funds brought in through their efforts. The dean, in this environment, is there in order to serve the professors, and they, in turn, are paid salaries to serve the junior faculty, who actually do the research for which the grants are made, and the students, who help in this work.

The principal usefulness of academic committees is in getting people to know each other. You get to know your colleagues very well indeed when you sit with them for a couple of hours once a week, to talk, say, about the grading of faculty by students. It is possible to learn more about the substance, the inner resources, and the reliability of a man or a woman in this way than by going off on a canoe trip in white water. You learn quickly who is to be trusted and who is to be worried about.

The main function of committees, and the one most likely to affect in an enduring way the quality and future destiny of the university, is the nomination of faculty members for promotion to tenure rank. This is the single area in which the dean can exert real power, since selecting the membership of tenure committees is, in most places, the prerogative of the dean. If he has an idea that Associate Professor Smith would be a likely prospect for a permanent appointment, good for the future name of the university, he can choose a search committee likely to come down on Smith's side—or at least he can avoid appointing members likely to hold prejudices against Smith or Smith's field. In some universities, the faculty are aware of the dean's power in this matter and mistrust him, and therefore insist that the tenure committee must be a standing committee not subject to a change in its membership by the dean's whim.

The principal task of the administrators—the president, the provost, and the deans—after making sure that the proper systems are in place for keeping track of the money and generating reliable reports to the outside world in accounting for all funds, is to Let Nature Take Its Course. At any rate, this is the main part of the job in an established university with a distinguished name and a good record. Not to meddle is the trick to be learned. The university is perhaps the greatest of all social inventions, a marvel of civilization, a product of collective human wisdom working at its best. A good university doesn't need to be headed as much as to be given its head, and it is the administrator's task—not at all an easy one—to see that this happens. The temptations to intervene from the top, to reach in and try to change the way the place works, to arrive at one's desk each morning with one's mind filled with exhilarating ideas for revitalizing the whole institution, are temptations of the devil and need resisting with all the strength of the administrator's character.

Hands off is the safest rule of thumb. The hands of trustees, the state legislature, the alumni, the federal granting bureaucracies, the national professional and educational societies, and most of all the administrators, must be held off, waving wildly from a distance maybe but never touching the mechanism. I would as soon take command of a platoon of scuba divers and swim into a coral reef with the notion of making improvements in its arrangements for living as I would undertake revision in the ecosystem of a university. It needs a lot of leaving alone, and a lot of spontaneous, natural evolution.

A medical school is an anomaly within a university and works differently, sometimes placing the whole institution at risk for its principles. Medical schools are constantly having hands laid on them, from all sides, hands carrying money or threatening to take money away, other hands twisting the head of this part of the university and turning it in a direction skewed to one side, aimed at immediate service to the community, also aimed in the direction of money, money carrying the guarantee of something tangible delivered quickly in return. This is not the habit of the universities. Not to say that universities do not seek money, they do so day and night, but not, generally, with the promise of a service or a product.

I have lived most of my professional life in one medical school after another, and have a deep affection and admiration for these institutions, but I can see that some things are wrong with them and are beginning to go wronger still. If I were the president of a major university I would not want to take on a medical school, and if it already had one, I would be lying awake nights trying to figure out ways to get rid of it.

At the beginning, having a medical school was no great responsibility for a university, no trouble at all, and nice for the prestige of the whole place. Medical schools were small affairs by today's standards, a hundred students or fewer per class: two years of basic biomedical science taught by faculty who usually added strength to the university's resources for scientific research and teaching, and a small and relatively inexpensive clinical faculty for the last two years, most of whom made their own living in private practice and cost the university nothing for their services. The teaching hospitals were autonomous institutions, supported by the local community, managed as separate corporate entities unrelated to the university and maintained either as voluntary or municipal (or county) institutions. Medicine was a solidly respected career, intellectually rewarding if not famous for being lucrative; the applicants for admission were adequate in number to fill the classes, but not much in excess of that number. The medical school was often located in another part of town—

sometimes in another, distant part of the state—from the rest of the university, which, most of the time, was oblivious to its existence.

The great change began in the years immediately after World War II, with the expansion of extramural research programs of the National Institutes of Health. During the mid-1950s I was a member of what was then referred to as the "Senior Council" of the NIH, the National Advisory Health Council, which reported directly to the Surgeon General of the U.S. Public Health Service and was supposed to set policy. We had the time of our lives. Everything seemed possible. The Congress was fascinated by the possibilities that lay ahead in medical research, and Senator Lister Hill and Congressman John Fogarty were powerful figures who had already started to build their legislative careers on medical science. The medical schools of the country were of a mind to begin expanding their scientific facilities, and there was money all around. Dr. Frederick Stone, executive secretary for the council, was a skilled and ambitious bureaucrat, and Dr. James Shannon, the director of NIH, knew exactly where he wanted NIH to go and how to lead it to its destiny, which involved strengthening the nation's capacity for medical science by building research into the daily, central, and essential functions of the American medical schools.

In retrospect, it can be seen that the expansion of NIH and the recruitment of medical faculties for implementing the national mission of NIH represented one of the most intelligent and imaginative acts of any government in history, and NIH itself became, principally as the result of Shannon's sheer force of will and capacity to plan ahead, the greatest research institution on earth. Only one thing went wrong, a mistake no one involved in the early years envisioned: research became more expensive than anyone could have guessed. While NIH selected for excellence and picked the strongest universities and their medical schools for the effort, it became at the same time the accepted idea that *every* faculty member of *every* medical school in the country must be a working scientist with a grant from NIH and a laboratory at his disposal. As an inevitable result, the merit system for recruiting and promoting faculty members in the medical schools would henceforth be determined, in large part or in whole, by research productivity and papers published.

With this stimulus, the emergence of the modern medical center, now known at some universities as the Health Sciences Center, or some such term, began. Today, these creations dominate the scene at many universities. They are typically located at or near the edge of the campus, immense structures built around the core of a huge hospital, swarming with clinics, diagnostic laboratories, special buildings for rehabilitation, mental health, retardation, geriatrics, heart disease, cancer, stroke, and any number of other categorical programs that have, at one time or another down the years, caught the interest of one or another congressional committee. The central hospital is usually designated as the "university hospital"; sometimes the university owns it outright, or otherwise has contractual arrangements which give the university the key right to designate its own faculty as members of the hospital's professional staff, usually with their incomes provided by the hospital.

Most of these new medical centers are of great value to the communities around their doors, and many of them can fairly be regarded as national, even international, resources for the most skilled and specialized health care to be found anywhere. There is no question as to their excellence—indeed, they have had the effect of raising the professional standards for medical and nursing care across the country.

The only question—and now it is a question causing anguish for both medical school and university administrators—concerns their relevance to the university mission. The question did not arise so often in the years when the medical centers were being built, or in the years when there seemed to be plenty of money to meet their costs of upkeep. But now, in the 1980s, with demands for retrenchment in all governmental programs and outcries everywhere against the rising costs of medical care, particularly hospital costs, the relationship between these hospitals and their parent medical schools, and of both to their fiscal guardians, the universities, is becoming increasingly strained and uncomfortable.

Meanwhile, during the past ten or fifteen years, the medical schools themselves have undergone a great expansion. Not only has the number of schools in the country increased by 50 percent, the number of medical students in many schools has doubled, or more than doubled. This happened during a period when both the federal government and many state legislatures believed that we were short of doctors, and the schools were paid by capitation, a sort of bounty, for each student added to the entering classes. Now, with the federal cutbacks already launched, including sharp reductions in low-interest student loans, many of the medical schools are barely escaping bankruptcy. As for the students already in the system, and those now planning to enter it, the cost of medical education is becoming so high that only those backed by affluent families can pay their way. The annual tuition alone in the medical schools of private universities is already close to $10,000 for most, and rising above $20,000 in some. The state schools are considerably cheaper, but their costs are also rising steeply.

The universities themselves are now at risk. Step by step, they have assumed—probably without anyone realizing the magnitude of each step—the ultimate control and ultimate responsibility for a large sector of the nation's health-care system. The annual budget of some medical schools matches or exceeds the operating budget of all the rest of the university. The rosters of tenured faculty and the commitment to graduate and postdoctoral education have become disproportionately greater in many medical schools. And now, with the sure prospect of reductions in the funds flowing from Washington to support the medical schools, it is the universities and the trustees who will have to decide where to make the inevitable cuts. Most universities live chancily from year to year, depending heavily on contributions from their alumni and philanthropic friends, sailing close to the wind. It is not within their conceivable resources to pick up deficits of any size, and the medical school deficits are soon likely to become of great size indeed.

Somehow or other, the medical centers will have to do a better job of sorting out their component parts. The medical school faculties carry responsibilities for teaching, research, and patient care, and are largely dependent on the hospitals for their income. As integral parts of the universities, the medical schools ought not to be in the business of running immense hospitals, any more than the law schools should be running the local court system or the business schools operating the town's major corporations and banks. The teaching hospitals cannot divorce themselves completely from the medical schools with which they are affiliated, but they should be recognized and supported by society for what they are—complex and costly institutions which are indispensable not only for the local community but for the whole country, some of them indeed for the whole world.

# 17
# Rheumatoid Arthritis and Mycoplasmas

The greatest difficulty in trying to reason your way scientifically through the problems of human disease is that there are so few solid facts to reason with. It is not a science like physics or even biology, where the data have been accumulating in great mounds and the problem is to sort through them and make the connections on which theory can be based. For most of this century—by far the most productive of technology in the history of medicine—clues have been found through analogies to known disease states in animals, sometimes only vaguely resembling the human disease in question.

In rheumatoid arthritis, the only comparable diseases that occur spontaneously in animals are the infections caused by mycoplasmas. In several species of domestic animals, most persuasively in swine, the joint lesions caused by mycoplasmas are indistinguishable in their microscopic details from those of rheumatoid arthritis in man.

Because of this solitary clue, many attempts have been made in laboratories around the world to cultivate mycoplasma from the joint fluid and tissues in rheumatoid arthritis, with essentially negative results. There are several scattered reports of positive cultures, but my laboratories have been unable to confirm any of these. The problem remains unsettled, although the possibility is by no means excluded by the considerable literature of failed attempts. The mycoplasmas are strange, fastidious creatures, and the growth media required by some of the known species include bizarre, unaccountable nutrients. Just in the last few years, new species of mycoplasma have been found in some of the diseases of plants and insects which have been under close study for a half century or longer. The yellowing death of thousands of Miami's palm trees is now known to be caused by microorganisms of this kind. It would not be a great surprise, therefore, if a still-unidentified mycoplasma were connected to rheumatoid arthritis in man.

There is another wisp of a clue: the observation that mycoplasmas as a class are peculiarly vulnerable to the action of gold salts. The mechanism of gold's action is still unexplained, but the fact of its effect is solid. Side by side is the equally solid observation, dating back fifty years, that rheumatoid arthritis is sometimes cured by injections of gold salts; thus far, gold is the only therapy generally recognized to cure this disease, although its use is limited by the frequency and severity of toxic reactions to the metal.

However, the morphologic resemblance to certain animal diseases is, by itself, not much to go by for making up one's mind, and most of the skilled, responsible investigators of arthritis remain commendably skeptical of its meaning. Meanwhile, a great deal of evidence has accumulated to indicate something gone grossly wrong with the immunologic system in the disease, and rheumatoid arthritis is now listed in some standard textbooks as a disorder of autoimmunity: the standing theory is that antibodies are formed against one's own tissues and the joint lesions are the outcome of this anomalous event. But this is also an unproved theory, and the field remains wide open for speculation. Hence, by the way, the great number of best-selling paperbacks on how to cure your arthritis by special diets, mineral baths, exercises, meditation, and various combinations of vitamins. It is an unsettled problem.

In this circumstance, anyone can become overwhelmed by his own true belief, and I confess here to mine. I am persuaded by the connection, thin as it is, to mycoplasma infection in animals, and by the gold story. I cannot count the hours that I have wasted in my own laboratory, during the last twenty years, concocting one baroque broth after another, trying to grow mycoplasmas from arthritic joint fluid, always with negative results. I have been obsessed with the possibility, unable to give it up. In any other circumstance, I suppose my behavior might be classed as paranoid. My laboratory notebooks contain an intermittent but endless harangue on this one topic; in between more or less respectable experiments on endotoxin, streptococci, papain, whatever, this business of mycoplasmas and arthritis keeps popping up, like King Charles's head. I cannot leave it alone.

I have shifted my ground somewhat in the last several years. Very well, perhaps it is wrong to believe so ardently in mycoplasmas, but what about bacterial L-forms? These are quite different microorganisms in their genetic origin, but morphologically indistinguishable from mycoplasmas. They can be made in the laboratory by simply stripping off the walls of living bacteria and then growing the denuded fragile creatures in special media designed to prevent them from bursting. The outcome, once they are adapted to growing in cultures, are colonies that cannot be told apart from classical mycoplasmas.

I began working on wall-less bacteria ten years ago in my laboratory at Yale, while I was professor of pathology and dean, and brought them along to Sloan-Kettering. I have never worked on a problem with so many attractive, unpublishable diversions. L-forms are, for my taste, enormously interesting beings, leading the investigator down one garden path after another. Here is just one example:

A good way to make L-forms from normal streptococci is to grow the bacteria on agar plates containing a gradient of penicillin. Penicillin acts by preventing the bacteria from synthesizing the constituents of their walls, and when the concentration of penicillin is exactly right, you can produce wall-less organisms that are not killed off but abruptly change their mode of colony formation, and their growth requirements, to those characteristic of mycoplasmas. They are called L-forms because they were first described at the Lister Institute, in London. After cultivating such organisms in penicillin for a while, one can wean them away from the antibiotic, and then they will live forever in their wall-less form.

A blind garden path opened in my laboratory one winter day when I had occasion, for other reasons, to administer penicillin to some guinea pigs, who then died, to my surprise, within three days. I then learned something already known to some of my colleagues in infectious disease: penicillin is lethal for guinea pigs. Indeed, if Florey and Fleming had been using guinea pigs for their initial experiments with penicillin, we might never have entered the antibiotic era. The lethal action of penicillin in this single animal species has never been satisfactorily explained.

It occurred to me that there might be a connection between this and the capacity of penicillin to convert normal bacteria to L-forms. Guinea pigs are known to harbor their own species of streptococci and pneumococci, usually as latent infections in their lymph nodes, but sometimes flaring up in epidemics to produce overwhelming lethal infections. Perhaps penicillin might be setting up a gradient of antibiotic in these animals, with colonies of L-forms possessing some new toxic property as the result. It seemed a good enough idea to pursue, anyway, and I ordered several dozen guinea pigs and began drawing up protocols for the experiment. By the time I got around to it, because of the usual delays in academic purchasing offices, it was the middle of March.

We injected thirty guinea pigs with the properly lethal doses of penicillin and made elaborate preparations for culturing the blood and tissues of these animals at the moment of death. On the third day all was ready, but none of the guinea pigs was dead, not even perceptibly ill. More penicillin was injected, in higher and higher doses, and the animals thrived on it.

We must have gotten the wrong strain of guinea pigs, we decided. Others were ordered in, from other breeders in Connecticut and New York, and we set things up all over again. By now it was late April. We launched the experiment, a much bigger and more orderly one, and therefore, as it turned out, a much more impressive dud. Not a guinea pig died, or even sickened.

We went back to the literature and found a few remarks, here and there, about variability of the phenomenon. One laboratory, in the Netherlands, reported that its colony of guinea pigs had become resistant to penicillin, and their work had therefore been terminated. But most of the papers described 90–100 percent deaths from penicillin, even in very small doses.

We dropped the matter then, and became busy with other affairs through the following summer and early fall. Then, for reasons I've forgotten, we decided to try it again, and lined up a dozen guinea pigs for penicillin. It was now early in November. This time, all the animals were dead in three days. We went ahead with more guinea pigs, batch after batch, all vulnerable to penicillin, and for the next five months the laboratory was awash in cultures designed for the isolation of L-forms; although our results were still negative, we were optimistic and enthusiastic about the predicted outcome. And then, late in March, everything ground to a standstill. Pencillin was no longer lethal. From that time on, every two weeks we established as a laboratory routine the injection of six guinea pigs with penicillin. No deaths occurred through the summer or early autumn; then deaths began again in November and continued through March, then stopped again.

We did this for three years running, never having a long enough period of penicillin deaths to finish our planned experiments, and never able to figure out a reason for the regular seasonal occurrence, never even able to write a sufficiently illuminating account of these events for publication. The observations are still there in our notebooks, stuck, unexplained, waiting for next winter.

Throughout all this time, in and out of the guinea pig experiments, my colleague Dorothy McGregor and I have continued trying to isolate L-forms from specimens of joint fluid and biopsies of synovial tissue from patients with rheumatoid arthritis, thoughtfully supplied by surgical colleagues at various hospitals in New York and New Haven. We are still at it, ten years after first having the notion that it was a good notion. Not once have we cultured anything resembling an L-form colony, but we have seen some things in the centrifuged pellets of joint fluid that had been incubated for twenty-four hours or so that look rather like L-forms, and we have grown bacteria called corynebacteria from most of the fluids and biopsies. These bacteria all look alike and behave alike, and they are not recoverable from joint fluid or synovial tissues from patients with other kinds of arthritis. It may turn out that the idea is right, after all. Perhaps there are L-forms in there, behaving like mycoplasmas, and perhaps they are the L-forms of corynebacteria, and perhaps they have something to do with the etiology of rheumatoid arthritis. I hope so, for it has been a long time on one problem.

In 1971 I was invited by the University of California at Davis to come for a two-week visit as Regents Professor. It was a wonderfully flattering invitation, involving such congenial and undemanding duties as a few lectures and several informal seminars, plus an opportunity to learn a lot at first hand about whatever research projects I happened to be interested in, and I accepted quickly. I would have accepted in any case, but the great attraction was the faculty issuing the invitation and the part of the university to which I would be assigned. It was the Veterinary School.

The different campuses of the University of California have each developed one or another particularly strong area, and Davis has for decades been distinguished for having one of the country's best schools of veterinary medicine, rivaled only by

Cornell, Iowa, and Pennsylvania. I had first become fascinated by this field because of the influence of Richard Shope. Shope became a good friend during our months in the same tent on Okinawa, and he used to tell me at length, during dark evenings, about the accomplishments of veterinary science and his deep respect for veterinarians. He was of course trained initially as a physician, but most of his experimental work had involved animal diseases (rabbit papilloma and swine influenza preoccupied much of his life), and he had many close friends and collaborators among animal doctors. He had an impressive list of honorary degrees, and the one he prized the most was a D.V.M. (hon.) from the University of Utrecht.

I was assigned an office with access to a small laboratory in the Davis Veterinary Sciences Center and allowed to roam. The undergraduates were a considerable surprise. As a teacher-physician I had thought of medical students as the very top of the line, and I was not prepared for young men and women as excellent as those Davis veterinary students. It saddens me to say so, but their intellectual quality and verve, their curiosity and skepticism, most of all the sheer fun they were having as students, made them a more interesting lot than I had been used to in medical school.

Watching the students and faculty on their rounds was another small shock. The animal clients were not, as I had rather expected, treated as interesting objects or technological problems to be solved. We rounded through the barnyard wards of sick cattle and horses, pens of ailing hogs, sheds containing scores of pet dogs and cats, cages of birds, even two locked wards for monkeys and chimpanzees, and all of these animals were known and recognized as individuals by the scholars and their professors. Moreover, they were handled with as high a level of affection and regard as I could wish for if I were bedded down in any New York City hospital.

Later on, in seminars with the students, I found that the competition for admission to the veterinary school at Davis was just as intense as for the medical school, maybe more so, but the reasons for competing were not the same. Individual students had different careers in mind—some were hoping to specialize in large farm animals, others in horse-breeding establishments, others in city-bound pet animals, a few were looking forward to research opportunities in the federal Department of Agriculture or faculty posts in veterinary colleges—but none of them seemed to have in mind high income or social status. They were there, having the best of times, because they liked animals.

One reason I had been invited was that I had been working with mycoplasmas, which are of special concern, scientifically and economically, within the veterinary world because of the formidable epidemics of lethal disease they produce: pleuropneumonia in cattle; arthritis in pigs, goats, and sheep; pneumonia and encephalitis in chickens, turkeys, and other birds; a long list of others. It is an odd anomaly that organisms with so wide a host range among animals, even extending to those Miami palm trees, seem to have, relatively anyway, so little interest in humans. One important lung disease, once known as virus pneumonia or primary atypical pneumonia, on which I had worked at the Rockefeller Institute, was finally proved to be caused by a mycoplasma now called M. *pneumoniae*, and several types of mycoplasmas are now known to be implicated in genital infections (perhaps also involved in the causation of spontaneous abortion), but except for these ills, human beings are not known to be prone to mycoplasma infection.

However, because some of the animal diseases do resemble, in the details of the pathologic lesions in joints and blood vessels, such human diseases as rheumatoid arthritis, lupus, inflammation of the arteries, and encephalitis, and because I happened to be one of the interested parties in the field, my invitation to Davis made some sense.

Professor Henry Adler and his avian disease group at Davis had been studying for some years a respiratory disease of poultry caused by a mycoplasma called M. *gallisepticum*. One variant of this organism, labeled the S-6 strain, was unique for its capacity to produce an unusual neurological disease in turkeys. When a suspension of S-6 mycoplasmas was injected into young turkey poults, the birds became comatose on the sixth or seventh day and then, within twelve hours, died. Death was caused by the selective destruction and occlusion of the arteries in the brain; the only other affected vessels in the body were those in the connective tissues around the joints. The arterial lesions were of special interest to me, for they looked very much like those in a human disease known as polyarteritis nodosa. I had begun work with this mycoplasma several years before, in the pathology department at Yale. Among other things, I had learned that the S-6 mycoplasma possesses a special and unique affinity for the arteries of the turkey's brain; mycoplasma antigen can be demonstrated in these vessels within a day or two of infection, lodged in dense masses beneath the lining cells and often extending through all the layers of the arterial wall. The organisms carry a neurotoxin of some sort (still to be chemically identified), which produces signs of extensive brain damage within less than an hour when they are injected intravenously. To be active, the toxin has to reach the brain by way of its blood vessels; when the same dose or larger doses of mycoplasmas are injected directly into the brain tissue itself, there is no evidence of toxicity. The species specificity of the toxin is remarkable: it affects *only* turkeys, not chickens or pigeons or ducks, not rats or mice or hamsters, and only young turkeys at that.

I had been working for several years before this with another type of mycoplasma, affecting mice and rats, called M. *neurolyticum*. This agent, a particularly fragile organism, hard to grow in any culture medium, elaborates another sort of neurotoxin which affects mice and rats but no other laboratory animal or bird. The toxin itself is extremely delicate and difficult to work with; it can be stored frozen, but if left around at room temperature for a couple of hours it becomes totally inactive. The brain lesions caused by the toxin do not involve the arteries; instead, there are widespread cystic cavities within and among the neurones in the cerebral cortex, rather like the so-called spongiform lesions seen in certain types of brain disease in humans.

Then there is that business of gold salts as a treatment for rheumatoid arthritis. A mycoplasma now known as M. *arthritides*, which produces extensive, chronic arthritis in mice and rats, was encountered in infected mice by Albert Sabin in the late 1930s, while he was engaged in an unrelated problem involving the serial passage of mouse tissue suspension from animal to animal. At that time, it was known that gold salts were therapeutically useful in the treatment of humans suffering from rheumatoid arthritis, so Sabin, on a mild flyer, treated his mice with gold. The effects were dramatic: joint swellings went down and the mycoplasmas vanished. Since that time, thanks to Sabin, animal infections by all forms of mycoplasma have been shown to respond to gold. It is an interesting but not particularly useful finding, since the mycoplasma diseases can be treated more cheaply and safely with antibiotics. But highly useful for purposes of speculation. It is another piece of indirect evidence for the possibility that rheumatoid arthritis may be caused by a mycoplasma. Parenthetically, in this series of chance observations, it is worth noting that the beneficial effect in arthritis was first observed by Forestier, a French clinician who was seeking a treatment for tuberculosis in the 1920s and decided to try various metal salts, gold among them. None of the tuberculous patients was improved, but several of them had rheumatoid arthritis as well, and these were relieved of their joint disease by gold.

M. *arthritides* infections in mice and rats remain an interesting disease model, but not as close to human arthritis as another disease, which occurs spontaneously in pigs, is readily transmissible from pig to pig (but not to any other species), and is caused by a species of mycoplasma found only in pigs.

These observations were the immediate and practical reasons for my laboratory's interest in mycoplasmas, but there was another reason, quite a different one, for my obsession with these organisms over so many years. To put it briefly, mycoplasmas are incredibly beautiful creatures. I first saw them through a microscope in the early 1960s, when someone sent me a culture of M. *pneumoniae* which I wanted to use in order to compare its antigens with those of a streptococcus that I had isolated from patients with primary atypical pneumonia (it was first recovered in the lung tissue of a patient named McGinnis and accordingly called the McGinnococcus, later formally designated as Streptococcus MG). As it turned out, there is some sharing of antigens between the mycoplasma and the streptococcus, but this is not what caught, and held, my eye. I learned what mycoplasma colonies look like when they are grown for a few days on clear agar.

The technique for staring at mycoplasmas is the simplest of things. You need an agar culture with colonies growing on it—the colonies are themselves so small that you must use a hand lens to be sure they are there. Then you cut away a block of the agar, place it upright on a glass slide, and place on top a thin glass coverslip which has been immersed beforehand in a lovely blue dye and then dried. The colonies pick up the stain within less than a minute and can thus be examined under the highest power of an ordinary light microscope.

The colonies are about 20 micrometers in diameter, about twice the size of a white blood cell. They have perfectly round centers, which stain a very deep-blue color, and around each center is a sort of halo, paler blue, shading off to a grayish-blue, vaguely outlined, circular perimeter.

That's all. I cannot say why they are so lovely to look at, but they are. The central, dark staining core is a solid mass of mycoplasmas that have somehow tunneled deep into the agar; the halo is produced by a thin layer of organisms growing on the surface. How the mycoplasmas at the center manage to grow with such force down into the agar is not known, but it is the characteristic and identifying structural feature of this class of organisms, and perhaps it is this geometric configuration, and the sense of something energetic going on in these quiet, motionless structures, that make them so pretty to look at. The color as well. Under the oil-immersion lens, magnified 1200 times, each organism can be seen as a bright-blue speck, nothing more, but when you look at the millions of such sharp blue specks massed all together in great clouds, it is like looking at life itself.

Under the electron microscope, sliced into thin sections by a diamond knife and magnified 50,000 times, the mycoplasma changes from a featureless blue speck to something as big as a house, and as busy-looking. There is no cell wall, like the rigid, thick boundary that envelops ordinary bacteria, only a delicate two-layer membrane. Inside are the ribosomes, from which the proteins needed for living are manufactured, and here and there, in a delicate meshwork visible between the ribosomes, are the strands of DNA carrying the instructions. Some mycoplasmas are very small and round with little knobs at one pole, others are long snaky filaments; some become enormously swollen, the size of a human leukocyte. Some have their own viruses inside, similar to the bacteriophage particles of ordinary bacteria. No one knows what the viruses are doing there; they are obviously lodgers of some sort, perhaps a messenger service for carrying genetic information from one set of mycoplasmas to another. Conceivably, the toxins of mycoplasmas may be coded by DNA brought in by a bacteriophage, as seems to be the case with diphtheria and streptococci, but this remains to be determined.

The mycoplasmas are unique among microbes. Taxonomically, they cannot be classed among either the bacteria or the viruses. Some investigators believe they are the descendants of ancient bacteria that long ago lost their cell walls. This is an attractive notion, since it is known that you can take the walls off bacteria without killing the organisms, and when this is done the wall-less bacteria grow in tiny colonies indistinguishable from mycoplasmas. However, there are enough differences between mycoplasma DNA and that of all bacteria thus far studied to make the relationship arguable, and the argument goes on.

For all their small size and fragile appearance, mycoplasmas are tough and extraordinarily adaptable. Tissue culture laboratories are plagued by them: suddenly a pure line of someone's carefully cultured, highly treasured mammalian cells will begin to disintegrate because of contamination by mycoplasmas, brought in from the laboratory air or from the breath of the investigator. Once in, they are nearly impossible to exterminate from the cultures: they take up residence within the surface coating of the tissue culture cells, or even deep inside, and manage thus to protect themselves against the antibiotics that would ordinarily kill them off. Mycoplasmas have been found living in the tissues of insects, and some plants are evidently infected in this way. They can adapt themselves to unusual environments: one species has been found living free in a smoldering coal-mine hillock, another flourishing at 80 degrees Centigrade in a hot spring.

You have to respect forms of life like these.

They are even proposed, from time to time, on wholly speculative grounds, as candidates for the ancestral precursors of nucleated cells. They are highly plastic, because of having no boundary beyond the unit membrane, and capable of enlarging to the size of a proper nucleated cell. Moreover, they have cholesterol incorporated into their membranes, like nucleated cells and unlike bacteria. It is imaginable that they may have served at one time as hosts for other prokaryotic organisms. Primitive microorganisms of one kind or another may have crossed the thin boundary and lodged inside more than a billion years ago, and these boarders may have evolved, how I cannot guess, into the nuclei, mitochondria, and chloroplasts of today's plants and animals. It is only a guess, anyone's guess, probably (but not necessarily) beyond experimental proof. But it is one more reason, in my view, for feeling affection for these dear and enchanting little beings.

# 18
# MSKCC: The Memorial Sloan-Kettering Cancer Center

The Memorial Sloan-Kettering Cancer Center, where I have worked for the last ten years, stands on most of two city blocks between York and First avenues, linked at the corner of Sixty-eighth and York to the Rockefeller University and New York Hospital–Cornell Medical Center across the street. Memorial Hospital is itself the largest and most comprehensive cancer facility in the world, and the most extraordinary hospital I have ever observed from the inside. It is at once the most specialized and most general of hospitals, dealing not only with the specific problem of this one disease but with the numberless other ailments that can afflict patients who have cancer, from newborn infants to people in their nineties.

I don't know how many times I've been asked across dinner tables by sympathetic strangers who know where I work: "How do you stand the place, all those deaths?"

A few years ago Hubert Humphrey was a patient, in for

treatment of the recurrent bladder cancer which eventually killed him, fully aware of the gravity of his situation, worried and somber on the evening of his admission, quiet and thoughtful, alone in his room at the time of my visit. We talked for a while; he was well-informed about his plight, knew that his chances for survival were slim, almost nil. But the "almost" was the focus of his attention. Over the next few days he transformed himself, I think quite deliberately, into the ebullient, enthusiastic, endlessly talkative Humphrey—not so much for his own sake as for what he saw around him. There were about forty patients on his floor, all with cancer of one type or another, some just arrived and undergoing diagnostic procedures in preparation for treatment; some due for surgery or chemotherapy next morning; others getting ready to go home with high hopes, cured once and for all; some at the end of the line, beginning to die.

Humphrey took on the whole floor as his new duty. Between his own trips to X ray or various other diagnostic units, he made ward rounds. He walked the wards in his bathrobe and slippers, stopping at every bedside for brief but exhilarating conversations, then ending up in the nurses' station, bringing all the nurses and interns to their feet smiling. During the several weeks he was in and out of the hospital, Humphrey's rounds became famous. One evening I saw him taking Gerald Ford along, introducing him delightedly as a brand-new friend for each of the patients. Together, Humphrey in his bathrobe and Ford in a dark-blue suit, nodding and smiling together, having a good time, Ford leaning down to be close to a sick patient's faint voice, they were the best of professionals, very high class.

It was an unusual event because of the eminence and conversational skill of Humphrey and his visitors, including especially Muriel Humphrey—one of the world's nicest women. But this sort of thing goes on all the time in Memorial. Patients who can be up and around are constantly circling the floor, finding new friends, stopping by the bedsides of others, making small talk. One hears, down the corridor, someone's voice saying, "Oh, I had that and it was nothing, really nothing. You just wait, you'll be feeling better."

The direct, day-to-day involvement of laymen in the creation and sustenance of institutions like Memorial Sloan-Kettering is a uniquely American phenomenon. The government had nothing to do with the original decision to build such a place, nor much to do with sponsoring it until the post-World War II years. The Rockefeller family was responsible at the outset. John D. Rockefeller, Jr., perceived, late in the nineteenth century, the need for combining a hospital specializing in cancer treatment with a research facility committed to this one disease, and assembled the land on which the present buildings were begun in the 1930s. His son Laurance has committed much of his time and energy to the whole center, serving as a chairman of the board for a quarter century, and enlisting help from other members of his family and a wide circle of friends to keep the endowment growing.

And then there are the volunteers, several hundred men and women of all ages, some of them healed patients or relatives of patients, some who simply came on the scene asking to help, and during summertime scores of teenagers swarming in and out of all parts of the center. These people do hard work, and a good many do it every day, part-time or all day long, whatever can be spared from their regular occupations. They push carts filled with books and magazines from room to room, bring flowers, run errands, help at the desks in the busiest parts of the outpatient clinics, and make conversation with the patients and their families whenever called upon for friendship, which is almost all the time. The fifteenth floor contains a commodious pavilion and a well-stocked patients' library, run by a skilled librarian and her volunteer associates. There is a piano, played any time by patients and ceremonially played several times weekly by professional performers together with other instrumentalists and singers who come in for the pure pleasure of entertaining patients with cancer.

The reason all this works is that in Memorial there are no minor problems, no small stakes. Some patients who come in with the diagnosis of cancer are discovered to have a lesser disease and go home elated. Others who have been operated on or given chemotherapy previously are admitted because of a suspected recurrence, and are vastly relieved to find that it isn't so. But all of the patients who enter Memorial do so in the full expectation that they do indeed have cancer, with no choice but to face up to it and have done whatever can be done. They, and their families and friends, are frightened by this disease as by no other, and they come to their rooms fearing pain and death, in need of all the reassurance they can find.

Sometimes the reassurance comes promptly and with solid scientific evidence. Women with early breast cancer and without lymph node spread can be surgically guaranteed a 95 percent prospect of lifetime survival without recurrence; for those with nodes the outlook is chancier, but the survival figure with combined surgery and chemotherapy stands now at around 70 percent. Lung cancer, perceived as incurable just a few years ago, seems now to be curable in as high as 40 percent of patients if their disease is spotted in its earliest stage, thanks in large part to the work of master surgeons like Edward Beattie and his colleagues. The malignancies of childhood, including leukemia, used to be the most rapidly lethal of all forms of cancer, but now, with combinations of various new drugs, they are becoming the most easily cured ("cure" is a word that has to be used provisionally, but many of these children have by now become young adults, healthy and free of any sign of disease). Testicular cancer, a malignancy of young men, was uniformly and rapidly fatal a short while ago; now, with new forms of chemotherapy, it appears to be curable in most patients.

The statistics show that in recent years approximately half of all Memorial patients are free of disease and in restored general health when discharged from treatment.

So, coming into the hospital is a sort of gamble, with 50–50 odds in favor of both the house and the patient. This is no comfort to the patients with advanced lung cancer, or those with widespread metastases from breast or colon or prostate cancers. These people know there is no real hope, and they need hope more than anything. What they can be given is a very small piece of hope: there are very few forms of cancer, even the most widespread and rapidly growing, that cannot be slowed down by chemotherapy. Months, as much as several years, can be added to productive life. For the elderly this brief extension of living comes as a boon, giving time for finishing any number of necessary tasks, setting affairs in order, preparing as we all must sooner or later for dying, but with time to concentrate on the preparation. For the younger patients the gain of a few months or even a few years seems much less of a reward, but they have something else to hope for: that in the meantime something new may emerge with the capacity to turn their cancer cells around. This, by the way, is an entirely rational prospect to have in mind. The pace of current biomedical research, particularly in the basic sciences relevant to cancer, has become so rapid, and the new insights so illuminating for the biology of normal and neoplastic cells, that it is reasonable now to hope for totally new therapeutic approaches, sooner rather than later.

Hope is itself a kind of medicine. I believe this, although I cannot prove it. I have not had a long personal experience in observing cancer patients, but some of my savviest colleagues who have been oncologists all their professional lives are convinced that patients who are able to maintain some sense of optimism do better, live longer, and recover more often than those who are discouraged from the outset and give up hope. There are a few studies published in the medical literature to back up this impression, but not enough data to make it solid. One paper by an Australian oncologist compares the mode of dying in patients who are convinced that they are doomed to the "bone-pointing" syndrome observed in aboriginal societies: when a witch doctor points a bone at a person, death follows within a few weeks. The great Harvard physiologist Walter Cannon once made a study of the hexing phenomenon and concluded that certain human beings do indeed slip into a state of apathy and die simply because of being told that this will happen. Neither Cannon nor anyone else has ever been able to guess how such a thing can happen, but happen it does, raising the possibility that the opposite sort of assurance given to genuinely sick people in the process of dying might have the effect of increasing resistance to the process. It may be because of this that so many charlatans have achieved temporary success in treating cancer. Their biologically inert pills or injections may seem to work wonders for a while, because of belief alone.

Cancer is generally thought to be a painful disease. Sometimes it is, especially in those cases where the bones and nerves become involved in the process, but the majority of patients with fatal cancer do not have pain. Indeed, dying from cancer is for most people a much less painful and a considerably more peaceful experience than death from other causes. When pain does occur, it can almost always be fully controlled. My colleagues who run the pain clinic at Memorial tell me that it can *always* be controlled by morphine or the newer derivatives of morphine, provided the physicians and nurses are skilled in drug administration and not resistant to using whatever doses are needed. The last thing to worry about in caring for a patient with terminal cancer is addiction, but some doctors and nurses do still worry about this, and their patients are subjected to unnecessary pain.

Up to now, most of the improvements in our technological capacity to deal with neoplastic diseases have arrived piecemeal, one by one, individualized for each of the different forms of cancer. The drugs that work so beautifully for childhood leukemia are not much help for adult patients with metastatic growths. Breast cancer can be dealt with by hormonal therapies that do not affect other tumors. Radiation treatment alone will eliminate certain cancers, while others are totally resistant. Because of such different responses it is believed in some quarters that cancer is really not a single disease but perhaps a hundred different ones, each requiring its own separate research program and, ultimately, its own special kind of treatment. Sometimes this rather bleak point of view is put forward defensively by the groups most concerned with public support for cancer research, within government and in the philanthropic foundations, in apprehension that public expectations of something easy and quick—a universal "answer to cancer"—are essentially unrealistic and may result in a backlash retreat from the necessary long-term commitment.

My own belief, based more on hunch than data, is that the notion is fundamentally wrong. In the end, when all the basic facts are in, I think it will turn out that all forms of cancer, in whatever organs and of whatever cell types, are a single disease, caused by a single, central controlling mechanism gone wrong. It is still too early to lay bets, but I would bet anyway that there is a gene or a set of genes in all cells, normally held under repression in healthy cells, which somehow escapes control and leads to cancer. The genes may be related to those switched on during early embryonic development, when rapid and almost unrestrained multiplication of cells is necessary, and then switched off when differentiated tissues have been perfected. There is already some evidence for such a mechanism, emerging from current research in molecular genetics. It has been proposed that the viruses known to cause cancer in experimental animals do so by carrying along within their own DNA some of the mammalian genes which release other genes responsible for unrestrained and undifferentiated growth. It will eventually turn out, I believe, that the chemical carcinogens act at the same locus within cells, switching on the same pathological mechanism that is involved in virus-induced tumors.

The idea that cancers in different organs represent separate, different diseases seems to me beyond belief even at today's level of knowledge. There are chemical carcinogens that will cause solitary cancers of the liver or kidney or lung or brain depending on the age of the animal, the dosage of the chemical, or relatively minor molecular modifications of the molecule. Viruses that produce cancer of one organ, the liver, say, in one animal species will cause cancer in the kidney or lung of others. A German investigator, Druckrey, showed long ago that a nitrosamine derivative that produces liver or lung tumors in adult rats will produce brain tumors when administered to fetal rats, but the brain cancer does not begin to grow until the fetus has reached late adult life.

All of which suggests, to my mind anyway, that cancer is still a problem for basic science. But progress in the related fields of cell biology, molecular genetics, and immunology has moved along so rapidly in just the last three or four years that it will not come as a great surprise to learn that there is, in fact, a single determining mechanism underlying all types of human cancer—although the nature of that mechanism is, of course, bound to be an astonishment. When it comes, this level of deep information will begin to generate pharmacologic ideas aimed at switching the mechanism off, or turning it around, and when this point is reached, we can begin talking about "a" cancer cure.

At another level, less profound but perhaps easier to get at, is the problem of resistance to cancer. It is remarkable that heavy cigarette smoking leads to lung cancer in as high as 10 percent of lifetime smokers, but even more remarkable that 90 percent do not develop cancer. The overall incidence of cancer of all types, in all societies, whatever the difference in environmental hazards, is estimated to be fixed at around 25 percent, suggesting that three quarters of us may be in possession of mechanisms for successfully resisting cancer throughout our lives. There are some cancer patients who seem to be cancer prone; multiple cancers affecting different organs are by no means rare, and the statistical probability that a patient who has been surgically cured of one type of cancer will subsequently develop another type in another tissue is significantly higher than for the population at large. Children born with defective immunological systems are much more likely to develop lymphomas than normal children.

It may be that there is an immune reaction to the appearance of the first cancer cells, which is mobilized as soon as the alien nature of these cells is recognized. Such a mechanism, if it exists in human cancer, might be the one that protects the 75 percent of us who will never develop the disease. Perhaps all of us are experiencing, from one carcinogenic environmental influence or another, the emergence of single cancer cells and a few of their progeny from time to time, in one tissue or another, and eliminating them promptly when they are perceived as foreign by our lymphocytes. If the recognition comes too late, or not at all, cancer develops—and that accounts for the susceptible 25 percent of us. I proposed this notion twenty-five years ago, and it was elaborated later by Burnet under the term "immunosurveillance." It remains an unproven theory, but I retain high hopes for it as well as a certain affection, since it was one of my few excursions into theoretical biology.

In recent years, the theory has received indirect but solid support from an unrelated quarter in medicine. A substantial number of people around the world have by this time received kidney or heart transplants, surviving successfully under prolonged treatment with drugs which suppress the cellular immunity system which would otherwise cause rejection of the grafts. Among these patients, approximately 10 percent have developed cancer within the first year after transplantation. Of the renal transplant patients who have survived for ten years, the cancer incidence is nearly 50 percent. The cancers have been of all varieties, but with a much higher percentage of lymphomas than would be expected in the age groups involved. One explanation for this phenomenon—the currently conventional one—is that the immunosuppressive drugs are themselves directly carcinogenic. The alternative explanation, which I favor, is that the emergence of these malignancies is the predictable result of the loss of "natural" immunity. The same tumors would be appearing in all the rest of us, except for our natural capacity to kill off the first cells every time.

The rare but spectacular phenomenon of spontaneous remission of cancer persists in the annals of medicine, totally inexplicable but real, a hypothetical straw to clutch in the search for cure. From time to time patients turn up with far advanced cancer, beyond the possibility of cure. They undergo exploratory surgery, the surgeon observes metastases throughout the peritoneal cavity and liver, and the patient is sent home to die, only to turn up again ten years later free of disease and in good health. There are now several hundred such cases in world scientific literature, and no one doubts the validity of the observations. But no one has the ghost of an idea how it happens. Some have suggested the sudden mobilization of immunological defense, others propose that an intervening infection by bacteria or viruses has done something to destroy the cancer cells, but no one knows. It is a fascinating mystery, but at the same time a solid basis for hope in the future: if several hundred patients have succeeded in doing this sort of thing, eliminating vast numbers of malignant cells on their own, the possibility that medicine can learn to accomplish the same thing at will is surely within reach of imagining.

I look for the end of cancer before this century is over. It used to be the convention for people in my position to guess fifty years out, just to be safe, but I have become much more optimistic in the last few years. Indeed, I now believe it could begin to fall in place at almost any time, starting next year or even next week, depending on the intensity, quality, and luck of basic research. The world of medicine is becoming filled with the prospect of surprises, and this will surely be one of them.

When it does come about, I of course hope it happens first at Memorial Sloan-Kettering. But it could happen anywhere in the world, and with the system for exchanging scientific information working as swiftly and accurately as it does these days, with the news of last week's experiments in Pasadena or New York or Paris reaching the ears of researchers in Tokyo or New York or Melbourne almost overnight, it will be a difficult task for scientific bibliographers and historians to sort out the credits. Indeed, one of the splendid features of the scientific enterprise has always been the urgency with which the participants have insisted on displaying the results of their work as soon as it is completed. With very few exceptions (most of these involving the technology of commercial developments for the marketplace rather than the items of actual discovery) there are no kept secrets in research. The only sure reward for the investigators is the exhibition of their work for everyone else to see. The published paper, ready for public scrutiny and criticism, is the whole point of the profession and the only way of advancement for the working scientist. There are no real national boundaries or barriers: Western immunologists know, down to the finest detail, what is happening in their field in Prague; Western mathematicians know what their colleagues in Warsaw and Lublin are up to; the theoretical physicists at Columbia seem to know, in general, what is going on in their fields in Moscow.

For cancer, the credits will have to be widely distributed no matter what institution claims priority for the final, successful step. I imagine that the last decisive answer, whatever it is, will come as an astonishment to everyone, and most investigators will then wonder why someone else thought of doing precisely the right experiment before they thought of it, but it cannot possibly come as an *overwhelming* surprise. It will almost certainly be a piece of new information that will fit perfectly, locking itself neatly in place at the apex of the huge mass of information already accumulated, and it will depend, for the sense it will make, on the preexisting coherence of that mass.

I shall rationalize in this way if the ultimate discovery turns out not to be made at Memorial Sloan-Kettering. I shall claim, and I am even inclined to claim it now, in advance, that Memorial Hospital has already been at it for almost a century, feeding in one item after another down through the years, and the Sloan-Kettering Institute has already invested over thirty-five years of its talent and energy on the problem, building a solid part of the pyramid now waiting to be topped off by *exactly* the right new set of experiments.

# 19
# Olfaction and the Tracking Mouse

I learned a little about olfaction during my residency in neurology at the Neurological Institute in the late 1930s. The former chief of neurosurgery, Dr. Joseph Elsberg, was still working at the institute then, and his scientific obsession was the use of olfactory acuity as a physical sign in the diagnosis of brain disease. He had worked out a complex system of glass vessels with blowers connected to tubes fitted into the nostrils with which he and his assistants were able to measure, with a rough degree of quantitation, the perception of tiny amounts of juniper, camphor, cinnamon, and the like in each side of the nose. The group had established the usefulness of the procedure in the localization of certain malignant brain tumors located in the deeper regions of the frontal and temporal lobes. The trouble with the procedure was that it required extreme patience and experience, and great skill, on the part of the technicians who carried it out, and the results were of significance in such a small proportion of the patients in the hospital that it was finally given up. None of the younger members of the neurosurgical staff was interested enough to continue the research work after Elsberg retired.

I did some reading in the field at the time, and found the literature on olfaction obscure and sparse. Nobody seemed to know much about the matter. The olfactory receptor cells were known to be bona fide brain cells, the only proper neurones in the brain that are exposed to the outside world and act as their own receptors of information from the environment. All the others, those concerned with the senses of touch, position, taste, hearing, and vision, depend upon relays of nerve impulses coming in from highly specialized receptor cells which pick up the appropriate stimuli at the periphery and send them along to centers in the interior of the brain which are set up to make sense of the sense. The most curious and remarkable thing about the olfactory neurones is that they come and go, replicating and replacing themselves in their positions at the surface of the olfactory mucosa, high up in the back of the nose. No other brain neurones have the property of multiplying or regenerating; once in place, the neurones of the rest of the brain are there for the duration of life, and those which age and die off are not replaced. But the olfactory receptor neurones keep coming; in the mouse they have been shown to have a turnover of the population every two to three weeks. Another peculiarity of these cells is that despite their exposure to the outside, and their location in a region of the air passages which is especially rich in bacteria and viruses of all sorts, the tissues in which they reside do not become infected. It was thought at one time that the poliomyelitis virus made its way into the brain by way of the olfactory neurones, but this was later proved wrong. It is now believed that the cells are somehow protected by the antimicrobial property of the mucus which is always present as a thin layer covering the cells.

From time to time, new information about olfaction appeared in the physiology literature, and a series of international conferences dealing with the phenomenon began in the 1950s. I kept in touch with this material as best I could, as an outsider, and one day about ten years ago I ran into some references to observations made much earlier—as far back as the 1920s—on the accomplishments of tracking hounds. A great deal of solid work had been done at that time, most of it sponsored by European police departments, on the capacity of trained dogs to track the footsteps of a single human being across fields marked at the same time by the tracks of other people.

Much of this was anecdotal, based on single observations made in the course of field trials and interdepartmental competitions, but the anecdotes were abundant and consistent, building a consensus accepted all round: a well-trained hound could distinguish with accuracy an odor of some kind arising from the track of a human being, for as long as forty-eight hours after the laying of the track, and could distinguish this particular track from all other tracks laid by other human beings.

If this was indeed true, it meant that the dog was able to smell a signal coming from the track which identified each human being as an individual self. But another elaborate and precise biological system was already known to exist for the same purpose: immunological markers that signal selfness are present at the surface of all cells in the body; the sensing of these chemical molecules is responsible for the fact that skin grafts are rejected with surgical precision when the grafts come from someone else, unless the foreign skin is taken from one identical twin and sewn into the skin of the other. This is believed to be a universal phenomenon: except for the exchange of tissues between identical twins, the skin of no one of the 4 billion human beings on the planet can be grafted successfully to any other. Grafting can be done these days with kidneys, even hearts, but only with the aid of drugs that incapacitate the lymphocytes responsible for immunologic rejection of non-self tissues.

It seemed to me strange that two different systems would have arisen for the same function in evolution, separately and unrelated to each other, and I began to speculate that they might indeed have evolved from a single ancestral system employed, early on in evolution, for enabling the first primitive organisms to make distinctions between their own cell surfaces and those of others. Such mechanisms are known to exist abundantly in the most ancient of metazoan creatures, sponges and corals for instance. Moscona showed some years ago that when the separated cells of two species of sponge are mixed together and rotated in a saline suspension, the two kinds of cells will reaggregate in clusters, each of which is made up exclusively of one or the other species. The cells can evidently recognize their own kind as self, and can at the same time avoid sticking to the non-self cells. Jacques Theodor placed together two pieces of soft coral of the same species, but from different colonies on the reef, and observed that, after first fusing to form a single frond of coral, the two bits would then separate from each other, precisely along the original line of apposition, with death of all the cells in the immediate vicinity of that line. Recently, Hildemann and his associates have observed the same sort of graft rejection in sponges: two sponges from the same colony will fuse together permanently, but when the sponges are of the same species but from different colonies, they will reject each other ten or twelve days after fusing. Moreover, the sponges seem to have a specific memory of the event; when the separated explants are again placed together, but with different surfaces confronting each other, the rejection reaction occurs in accelerated fashion, two to three days later. The phenomenon is remarkably like graft rejection in the mouse: the first skin graft from a foreign mouse is rejected in eight to ten days, but a second graft of the same tissue is rejected in three or four days.

In mice, the reaction of graft rejection is primarily a function of a particular class of lymphocytes, the so-called T-lymphocytes (so designated because of their origin in the thymus gland). The reaction is controlled by a special set of genes, called the H-2 locus, always situated together on a single chromosome. In man, the corresponding gene locus, governing the self–non-self distinction and the rejection of tissue grafts, is known as the HLA locus.

Several years ago, when invited to deliver an address before an Immunology Congress on possible future lines of immunologic research, I discussed the problem of self-marking and expressed the view that it would be remarkably unparsimonious of nature to set up two such elaborate and complex systems for individual self-marking—costly in terms of energy, one involving the immunologic markers of histocompatibility and the other using olfaction—and to have these two mechanisms evolving without being closely related to each other. I made at that time what I thought was a mild biological joke, predicting that the same set of genes would be found responsible for both systems of labeling, and that someday "man's best friend would be used for sniffing out histocompatible donors."

A while ago I was discussing this with Dr. Edward Boyse, whose research laboratory in Sloan-Kettering makes daily use of an extensive collection of meticulously inbred and sharply defined lines of mice. His wife, Jeanette Boyse, has the immediate responsibility for overseeing the breeding of various lines of congenic mice, in which the sole genetic difference between two lines lies in the H-2 locus on chromosome 17, the locus governing graft rejection and coding out the major histocompatibility complex (the MHC) of tissue antigens. The mice are contained in transparent boxes so that their mating behavior can be kept under close and constant observation. Mrs. Boyse had just noticed that the males of certain lines displayed a preference for mating with females of the opposite line possessing different H-2 genes. The possibility was raised that perhaps the male could smell the difference, and since these were two lines of genetically identical animals, except for the H-2 difference, it was obvious that the capacity of a male to smell such a difference would have to involve an olfactory distinction between self (in strict terms of individual self) and non-self.

It did not take long for the Boyses, together with two young postdoctoral fellows at Sloan-Kettering, Drs. Yamazaki and Yamaguchi, to establish with satisfactory statistical significance that the phenomenon of mating preference between H-2 congenic mice was real and consistent. We then moved on to a simpler system for getting at the same problem, which involved, at the outset, training a tracking mouse.

In brief, the technique was based on the classical Y-maze, with two different odors coming down the arms of the Y, one from the tracking mouse's own line, the other from the congenic line differing from himself only at the H-2 locus. The reward for selecting the correct arm was a drop of water, and the tracker was urged to seek the drop by being deprived of water for the preceding twenty-four hours. Training was begun by teaching the mouse to distinguish between the odor of cinnamon and juniper; then, when he'd got the idea, he was trained to discriminate between the smell of his own and a totally different breed of mouse, and finally to detect the odor of the congenic line, in this case the difference between B-6, his own line, and B-6 H-$2^k$, the other strain.

The experiments worked, and have continued to work, with a surprising degree of consistency and reproducibility. In all, eight smart tracking mice have been taught to smell H-2 during the past two years. Each experimental trial involves twenty-four runs toward the target, which is changed from one arm to the other at random, and the correct or incorrect choices are recorded by a third party who is himself unaware of the correctness of the choice. With a very high degree of statistical significance, each tracker has learned to distinguish between his own smell and the congenic smell when the odor box leading to the arm of the Y maze contains mice of the proper genetic line. The odor is not detectable in homogenates of various mouse tissues, including spleen, liver, kidney, lung, or brain, nor can it be detected when mouse embryos are in the box. However, it is readily detected, with an accuracy even greater than when whole live mice are in the odor box, in samples of urine. The tracker can detect the odor of congenic urine when the urine is contained in a petri dish in the odor box, and the smell is still perceived when the urine has been diluted 1–40. The odorant is surprisingly stable, withstanding boiling for one hour. It is a small enough molecule to pass through a dialysis sac.

We have since learned that the same odorant can be detected in the urine of F-2 segregants derived from crosses between the two congenic lines, effectively ruling out smells derived from parental environment or family litter boxes.

It would be very nice to know which cells in the body are responsible for manufacturing the self-identifying odorant that ends up in the urine. The leading candidates for this role, in my view, are the lymphocytes themselves, because of the central role they play in mediating the homograft-rejection mechanism. An experiment still in progress at the Monell Institute has already produced suggestive evidence in favor of this idea. Yamazaki has transformed a mouse of one congenic line (B-6) into one of the other line (B-6 H-$2^k$), by irradiating B-6 mice to destroy all of their bone marrow cells and then transfusing them with lymphoid cells from the H-$2^k$ line to repopulate the now-empty bone marrow. Thus, what used to be a pure-line B mouse was turned into a mouse with the K marker on its lymphocytes. Urine from such transformed mice was then tested in the Y-maze to see what it smelled like to the trained tracker. It was recognized as K urine, indicating that the odorant is secreted into the urine (and probably concentrated there) as the result of activity on the part of bone marrow cells—most likely the lymphocytes in the bone marrow.

The same odor is responsible for the phenomenon of pregnancy blocking, the so-called Bruce effect. This is the peculiar reaction that occurs when a newly impregnated mouse is placed in contact with a strange male: the pregnancy is promptly terminated and the female goes into estrus. She does not do so, of course, when the contact is with the original stud responsible for her pregnancy.

Using congenic lines of mice, differing only at H-2, Yamazaki and the group have found that replacing the original stud with a different male of the same line does not cause the

Bruce effect, but when the new male is of the line with a different H-2 locus, the pregnancy stops and estrus resumes in the majority of females. The actual presence of the H-2 foreign male is not needed for this effect; the same results occur when the pregnant female is in the immediate vicinity of a sample of urine from the appropriate line.

The Bruce effect is not induced by exposure to *females* of the congenic line, or by urine from such females. Thus the Bruce effect must be caused by the perception by the pregnant female of two distinct and separate signals, one indicating maleness, the other announcing the presence of a male with a different H-2 locus.

I know of no satisfactory explanation for the Bruce effect, not anyway in teleological terms. Perhaps it represents a built-in response which tends to enhance heterozygosity and, to some extent, to impair close inbreeding. Or perhaps—and this is the teleology I prefer—the mere presence nearby of a strange male, differing in the odor of his H-2 from the original stud, signifies the departure of the father and the loss of protection to be expected from him for the forthcoming litter, and therefore it is time for the female to give it up and start over again. Our experiments have told us nothing about this, only that the smell of male strangeness is coded by the same string of genes that code for immunological strangeness.

Last year, Dr. Boyse and I had the opportunity of observing tracking dogs at work, first at the dog-training station of the Baltimore police department, then later at the Scotland Yard station just south of London. We saw enough to convince us that the specific and selective tracking of a man was a genuine and reproducible phenomenon, and that it ought to be entirely feasible to set up experiments to settle the question of whether identical twins leave identical tracks and even—although here I can envisage some formidably difficult technicalities—trying to correlate tracking accuracy with human HLA types. Needless to say, we have not set out on either of these lines, but some of the things we have already observed are perhaps of anecdotal interest even if not of scientific value. One curious thing we hadn't known: when a hound sets off on the track of a designated man, he does so not with his nose close to the ground, as in the movies, but rather tossing his head high, from side to side, as he goes. When the track turns at a sharp angle he overruns it, of course, but when he comes back to regain it he does so by sniffing the air well above the ground surface, getting clues not from the ground itself, or from footprints, but from something rising away from the ground.

In the Scotland Yard trials, we brought along several squares of gauze that had been placed in the bedding of various cages containing the two congenic groups of mice, differing from each other by only the genes for immunologic self-marking, and asked the trainer to see if his dog could learn to distinguish between the two. The squares were laid out at random, at intervals over a long slab of grass, the dog was given the scent of the one to be selected, and he trotted rapidly along with his head held several inches above the gauze squares until he reached the correct one, which he picked up neatly in his teeth and brought back to his master as though carrying the evening paper. The whole operation seemed so effortless as to be nearly automatic and, from the dog's point of view, the easiest of things. If we humans possess pheromones that label each of us as a person, I am glad to say that we cannot, as a rule anyway, smell them, social life being complicated enough as it is, but it would not surprise me at all if a Scotland Yard hound could do so, and could readily pick up the fragrance of any one urine sample and tell it from all the rest.

But even with the technical limitations of the tracking mouse and the Y-maze, it ought to become possible to learn something now about the chemical nature of the H-2 coded olfactant in mouse urine. Indeed, Drs. Yamazaki and Yamaguchi have recently transferred their laboratories to the Monell Institute in Philadelphia, where the chemistry of odorants is a high-technology specialty, for this purpose. There will surely be some interesting questions. What sort of heat-stable substance can it be, possessing enough variability in its structure to provide unique self-markers for the numberless individual mice, or, for that matter, 4 billion human beings? I imagine that it will turn out to be a *set* of chemicals, probably of the same class but with structural variations, arranged in infinite numbers of possible medleys, possibly very small changes in the intensity of one or another member of the group, and with each individual's odor sounding a unique chord.

Perhaps some similar arrangement of groups of molecular signals will account for the apparently infinite variability of cell markers in the immunologic system. It is conceivable that the tissue antigens are similar sets of different signals, displayed in varying concentrations to achieve uniqueness. It is not beyond imagining that the actual molecular configurations that fire off the olfactory receptor cells might turn out to be the same, or closely related to, the ones that, in the end, fire off a T-lymphocyte. And, to carry the matter as far as it can be stretched, it is even imaginable that some signal arrangement of this sort is at work in the homing of embryonic cells, the self-preservation of sponges, and the preservation of internal privacy within an amoeba. If so, it adds something more to the complexity of life for the single cell. It is not a simple life to be a single cell, although I have no right to say so, having been a single cell so long ago myself that I have no memory at all of that stage of my life.

# 20
# *Illness*

One of the hard things to learn in medicine, even harder to teach, is what it feels like to be a patient. In the old days, when serious illness was a more commonplace experience, shared round by everyone, the doctor had usually been through at least a few personal episodes on his own and had a pretty good idea of what it was like for his patient. A good many of the specialists in pulmonary disease who were brought up in the early years of this century had first acquired their interest in the field from having had tuberculosis themselves. Some of the leading figures in rehabilitation medicine had been crippled by poliomyelitis. And all physicians of those generations knew about pneumonia and typhoid at first hand, or at least once removed, in themselves or their immediate families.

It is very different today. The killing or near-killing illnesses are largely reserved for one's advancing years. No one goes through the six or eight perilous weeks of typhoid anymore, coming within sight of dying every day, getting through at the end with a stronger character perhaps, certainly with a different way of looking at life. The high technologies which are turned on to cope with serious disease—the electronic monitors in intensive care units, the chemotherapy drugs for cancer, the *tour de force* accomplishments of contemporary surgery, and the mobilization of increasingly complex procedures for diagnosis in medicine—are matters to be mastered only from lecture notes and books, and then by actual practice on patients, but very few doctors have more than an inkling of what it is actually like to go through such experiences. Even the childhood contagions are mostly gone, thanks to vaccines for measles, whooping cough, chicken pox, and the like, thanks especially to the easy control of streptococcal infections. Today's young doctors do not know what it is to have an earache, much less what it means to have an eardrum punctured.

The nearest thing to a personal education in illness is the grippe. It is almost all we have left in the way of on-the-job training, and I hope that somehow it can be spared as we proceed to eliminate so many other human diseases. Indeed, I would favor hanging on to grippe, and its cousin the common cold, for as long as possible. A case could be made, I think, for viewing the various viruses involved in these minor but impressive illnesses as a set of endangered species, essentially *good* for the human environment, something like snail darters.

Most people afflicted with grippe complain about it, and that is one of its virtues. It is a good thing for people to have, from time to time, something real to complain about, a genuine demon. It is also a good thing to be laid up once in a while, compelled by nature to stop doing whatever else and to take to bed. It is an especially good thing to have a fever and the malaise that goes along with fever, when you know that it will be gone in three or four days but meanwhile entitles you to all the privileges of the sick: bed rest, ice water on the bed table, aspirin, maybe an ice bag on the head or behind the neck, and the attentions of one's solicitous family. Sympathy: how many other opportunities turn up in a lifetime to engage the sympathy and concern of others for something that is not your fault and will surely be gone in a few days? Preserve the grippe, I say, and find some way to insert it into the practical curriculum of all medical students. Twice a year, say, the lecture hall in molecular biochemistry should be exposed to a silent aerosol of adenovirus, so that the whole class comes down at once. Schedules being what they are in medical school, this will assure that a good many students will be obliged to stay on their feet, working through the next days and nights with their muscle pains and fever, and learning what it is like not to be cared for. Good for them, and in a minor way good for their future as doctors.

The real problem is the shock of severe, dangerous illness, its unexpectedness and surprise. Most of us, patients and doctors alike, can ride almost all the way through life with no experience of real peril, and when it does come, it seems an outrage, a piece of unfairness. We are not used to disease as we used to be, and we are not at all used to being incorporated into a high technology.

I have learned something about this, but only recently, too late to do much for my skill at the bedside. On several occasions, starting around age sixty-four, I have had a close look from the bed itself at medicine and surgery and, as I shall relate, an even closer look at myself. On balance, I have very much liked what I have seen, but only in retrospect, once out of bed and home free. While there, I discovered that being a patient is hard work.

It is often said that people who have been precariously ill, especially those who have gone through surgical operations, love to talk about their trials and will do so at length to anyone ready to listen. I rather doubt this. Being ill is a peculiarly private experience, and most of the people I know who have gone through something serious tend to be reserved about it, changing the subject when it comes up. But here I am, about to talk about my times on the line and the things I learned. I only do so, I must say in advance, out of professional interest.

The first, and most surprising of all, was an obscure kind of pneumonia, chills, fever, prostration, and all, occurring suddenly on a Tuesday afternoon. I took to bed at home in good cheer, anticipating several days of warm soup, cold drinks, fluffed-up pillows, and ample family ministrations. But a week went by and I kept on with the chills and fever, so my wife called the doctor, a friend of mine and a real, house-calling doctor. He did the usual things, including taking samples of blood, murmured something about a virus "going around," and predicted that the fever would be gone in another day or two. But the next day I was in a hospital bed having more blood tests, being examined by platoons of interns and residents, and in and out of the X-ray department having pictures taken of all sites including bones. The laboratory tests had revealed a hemoglobin level of just under 8 grams percent, half the normal value, and it had become an urgent matter to discover where the blood had gone, or was going.

Within the next few days the pneumonia vanished, along with the chills and fever, and I had become a new sort of diagnostic problem.

To be worked up for anemia of unknown origin is strenuous exercise. The likeliest cause was blood loss, and the likeliest source was the intestinal tract—what is known, ominously, as silent bleeding. I received two transfusions, and then plunged headlong into technology. A bone marrow biopsy was, as I recall, the first piece of work, done neatly and quickly on a pelvic bone with rapid-fire explanations by the hematologist as he went along, telling me what I would feel and when it would hurt, but despite his reassurances I could not avoid the strong sense that having one's bone marrow sucked into a syringe was an unnatural act, no way for a human being to be treated. It did not in fact hurt much, but the small crunching of bone by the trocar followed by the peculiar and unfamiliar pain in the marrow itself were strange sensations, not at all nice.

I have performed bone marrow biopsies myself, long ago as an intern and from time to time since, and have always regarded the procedure as a minor one, almost painless, but it had never crossed my mind that it was, painless or not, so fundamentally unpleasant.

The rest of the workup was easy going, and at times engrossing. The walls of my stomach and upper intestinal tract were marvelously revealed by a barium meal, and those of my bowel by a similar enema, and all was well. But I continued to bleed, somewhere in that long channel, and more transfusions were needed.

It is not an easy task for doctors to look after doctors, and especially difficult when the doctor-patient is a colleague and close friend. It requires walking a fine line, making sure not to offend professional pride by talking down to the doctor-patient, but also making sure that the patient does what he is told to do. I was treated with great tact and firmness. The colleagues and friends who had me as a responsibility remained my good friends, but there was never a question as to my status: I was a patient and they expected me to behave like one. I was not to try making decisions about my own diagnosis and treatment. I was not allowed to go home for a few days, which I wanted very much to do at several times during what turned out to be a long period of hospitalization; it was explained very gently that the source of bleeding was still unknown and I might have a more massive hemorrhage at any time, rather less gently that if it happened I might suffer a lot of brain damage and I'd better not be doing that at home.

With negative X rays, my intestinal tract needed a different kind of look. On the possibility that I might have a polyp somewhere, bleeding freely but too small to show up by X ray, I was wheeled off to the endoscopy service for examination by the colonoscope, an incredibly long and flexible quartz fiber-optic tube through which all parts of the large intestine can be viewed under direct illumination by light sent in from outside. As a nice gesture of professional courtesy, the doctor stopped at frequent intervals during this procedure and passed the viewing end of the instrument over my shoulder and in front of my left eye. "Care to take a look?" he asked. I had never looked through this wonderful instrument before, although I had seen many photographs of the views to be had. It would have been interesting in any case, I suppose, but since it was the deep interior of my own intestine that I was looking at, I became totally absorbed. "What's that?" I cried, as something red moved into view. He took a look and said, "That's just you. Normal mucosa."

A few days after this fascinating but negative excursion, I had another episode of bleeding, my hemoglobin dropped to a disturbing level, more transfusions were given, and it was decided that I would probably need surgery in order to remove the part that was presumably bleeding. But without knowing the exact source, this could mean taking out a lot of intestine and, even then, missing it. The gastroenterologist who had me in charge, Dr. Paul Sherlock, knew of one obscure possibility not yet excluded, one that I had never heard about—a condition produced by an abnormal connection between an artery and vein in the intestinal wall—which had recently been reported as a cause of intestinal hemorrhage.

This was not, as it seemed to me at the time, a guess in the dark. The X rays and colonoscope had ruled out cancer of the colon (which is what I was pretty sure I had, at the outset), and diverticulosis (little cracks in the intestinal wall), and polyps had been excluded as well. The new syndrome of arteriovenous anomaly was about all that was left.

Finding out required close collaboration between the gastroenterologists and radiologists. A catheter was inserted in the femoral artery, high up in the right leg, and pushed up into the aorta until its tip reached the level of the main arteries branching off to supply the large intestine. At this point, an opaque dye was injected, to fill all those arteries. Just before pressing the syringe, Dr. Robin Watson, the X-ray chief, warned me that I would feel a sense of heat, not to worry. It was a brand-new sensory impression, perhaps never experienced except by patients undergoing this kind of arteriography: for about thirty seconds I felt as if the lower half of my body had suddenly caught fire, then the feeling was gone. Meanwhile, movies were being taken of the entire vascular bed reached by the dye, and the diagnosis was solidly confirmed. Dr. Watson came into the room a few minutes later with sample pictures displaying the lesion. "Care to take a look?" he asked. I was enchanted: there, in just one spot somewhere on the right side of my colon, was a spilled blur of dye, and the issue was settled. It struck me as a masterpiece of technological precision, also as a picture with a certain aesthetic quality, nice to look at. I could hear in the distance the voices of other doctors, quietly celebratory as doctors are when a difficult diagnosis is finally nailed down.

That evening I was visited by the anaesthesiologist for the brief but always reassuring explanation of the next day's events, and in the morning I rolled down the hall, into the elevator, and down to the operating room, pleasantly stoned from Valium. The next things I saw were the clock on the wall of the recovery room and the agreeable face of my friend the nurse in charge, who told me that it had gone very well. I have no memory of the operating room at all, only the sound of the wall switch and the hissing of the automatic doors as my litter entered the place.

That was my first personal experience with the kind of illness requiring hospital technology. Thinking back, I cannot find anything about it that I would want to change or try to improve, although it was indeed, parts of it anyway, like being launched personless on the assembly line of a great (but quiet) factory. I was indeed handled as an object needing close scrutiny and intricate fixing, procedure after procedure, test after test, carted from one part of the hospital to another day after day until the thing was settled. While it was going on I felt less like a human in trouble and more like a scientific problem to be solved as quickly as possible. What made it work, and kept such notions as "depersonalization" and "dehumanization" from even popping into my mind, was the absolute confidence I felt in the skill and intelligence of the people who had hold of me. In part this came from my own knowledge, beforehand, of their skill, but in larger part my confidence resulted from observing, as they went about their work, their own total confidence in themselves.

The next two course offerings were trauma. I'd never had real trauma before in all my life—or only once, when as a small boy in grammar school I'd been hit on the head by a pitched baseball and knocked out for a moment. The next day my head began to itch, and I went around the house and back and forth to school, scratching my head incessantly and complaining to everyone that the baseball had injured the nerves. My mother was skeptical of this, as she was about most self-diagnosis, and took a close look at my scalp. I was infested by dense families of head lice, caught perhaps from somebody's cap at school. In no time, but with a great deal of anguish caused by neat kerosene and larkspur shampoos, endless rakings with a fine-tooth comb, and finally a very short haircut, I was cured, although I'm not sure my mother ever thought of me again as quite the same boy. Anyway, that was all my trauma until I was sixty-six years old.

I was in the surf at Amagansett, floating in a high leisurely wave, and turned to catch it for a clean ride to the beach, when suddenly something went wrong with my right knee. After some floundering and swallowing a lot of water, I made the trip in under breakers and tried to stand. I couldn't, and had to wave for help. I was hoisted out and horsed along by friends, unable to place my weight on my right leg. On the way up the beach I was met by Dr. Herbert Chasis, a Bellevue colleague, and his wife, Barbara. Herbert knows more about kidney disease and hypertension that anyone I know, and Barbara had served as chief of the psychomedical wards during my years at Bellevue. They were sympathetic of course, but also to my surprise highly knowledgeable about knees and, like all good Bellevue people, ready to help. Indeed, Chasis had in his beach bag a proper knee brace, which he helped me put on, and explained to me that I had undoubtedly torn a knee cartilage (which he had done some years earlier; hence the equipment). Presently, crutches were provided as well, and off I went, home to get dressed and into Manhattan for a visit to the Hospital for Special Surgery. X rays again, and to my astonishment, another fiberoptics instrument, this time an arthroscope for looking into the interior of joints. In it went, moved around this part and that of the knee, and then it stopped. "Care to take a look?" asked the surgeon, handing me the eyepiece. I stared, transfixed, at the neat geometry of cartilage lining the joint, gray and glistening in the light, and then I saw what he had seen, a sizable piece of cartilage broken and dislodged. "Thank you," I thought to say, "what now?" "Out," he replied. So, Valium again, the unremembered operating room, and a long elliptical incision stitched with countless neat threads which I saw directly the next morning when the dressing was changed and I was commanded to get out of bed and stand on the leg, pain or no pain. Good for it, I was told, but I forget why. Teach it a lesson, maybe. Then on crutches for a few weeks, a single crutch for a few more, exercises thrice daily—lifting heavy weights by my foot (which I did for a few days and then began telling lies about)—and finally full recovery except for the odd pain now and then. Another triumph. I began to feel almost ready to write a textbook.

One more, and then I've finished, I trust. I was invited to give an evening lecture at the Cosmopolitan Club in New York, a place filled with dignity and intellectual women, including my wife. I prepared a talk on symbiosis, with a few lantern slides to illustrate one of my favorite models of insect behavior, the mimosa girdler. Halfway through the talk I called for the slides, the lights went out, I approached the screen to point to the location of the mandible-work of the beetle—and fell off the platform into the dark. Hauled to my feet, I found myself unable to move my left arm because of pain in the shoulder. I felt for the shoulder with my right hand and found an empty cavity. Someone brought a chair and sat me down while I caught my breath, and several sympathetic voices suggested that I cancel the evening. I wanted to finish the girdler story, however, and hunched back to the podium. It must have been a painful talk for the audience. I thought no one would notice the shoulder, and I droned on to finish the lecture, sweating into my eyeglasses and onto my manuscript, bracing my elbow against the podium, thinking that I was getting away with it nicely. Then I noticed that Dr. André Cournand, the great cardiologist from Bellevue, had moved from his seat in the front row and taken a chair alongside the podium, watching me carefully. I remembered the old anecdotes about that hospital, the car crash, the exploded street, and the figure of authority thrusting through the crowd: "Stand aside, I am a Bellevue man." If I topple, André will catch me, I thought.

Anyway, I finished it, although I'm not sure anyone was listening. An ambulance had been summoned, and I was carted off once more, on a litter, to Memorial. An X ray showed the shoulder dislocated and fractured ("Care to take a look?"), and my friends set about trying to get it back in place, me lying on my stomach buzzing with morphine, with the left arm hanging down, Ted Beattie (chief of surgery at Memorial) on the floor pulling down on my arm, but it wouldn't slip back in place. Finally, after a half hour of tugging, it was decided that I would have to be fixed under a general anaesthetic, and a call went out for the head of the anaesthesia department, Paul Goldiner. It was now 10 p.m., and Paul was at home in Westchester County. He hurried to his car and started the drive along the parkway to New York. Five minutes from home he passed the scene of a car crash and noticed a man lying prone on the grass. He stopped, went over to the man, found him pulseless and not breathing, and set about resuscitation. Anaesthesiologists are the best of all professionals at this skill, and shortly he had the man breathing again and conscious. He ran back to his car and set off again for New York. By this time, Beattie's maneuvers had snapped my humerus back into its joint, but it was too late to call Goldiner and tell him to stay home. He arrived at the door to be informed that his trip had been unnecessary, then told his own jubilant story; the trip had been the best he'd ever made. It is not often given to chiefs of anaesthesiology to save a life on the open highway, and Goldiner was a happy man. I have often wished I could locate his patient someday and have him take Paul Goldiner and me out to lunch.

That was the last of my trial runs as patient, up to now anyway. I know a lot more than I used to know about hospitals, medicine, nurses, and doctors, and I am more than ever a believer in the usefulness of technology, the higher the better. But I wish there were some easier way to come by this level of comprehension for medical students and interns, maybe a way of setting up electronic models like the simulated aircraft coming in for crash landings used for pilot training. Every young doctor should know exactly what it is like to have things go catastrophically wrong, and to be personally mortal. It makes for a better practice.

I have seen a lot of my inner self, more than most people, and you'd think I would have gained some new insight, even some sense of illumination, but I am as much in the dark as ever. I do not feel connected to myself in any new way. Indeed, if anything, the distance seems to have increased, and I am personally more a dualism than ever, made up of structure after structure over which I have no say at all. I have the feeling now that if I were to keep at it, looking everywhere with lenses and bright lights, even into the ventricles of my brain (which is a technical feasibility if I wanted to try it), or inside the arteries of the heart (another easy technique these days), I would be brought no closer to myself. I exist, I'm sure of that, but not in the midst of all that soft machinery. If I am, as I suppose is the case at bottom, an assemblage of electromagnetic particles, I now doubt that there is any center, any passenger compartment, any private green room where I am to be found in residence. I conclude that the arrangement runs itself, beyond my management, needing repairs by experts from time to time, but by and large running well, and I am glad I don't have to worry about the details. If I were really at the controls, in full charge, keeping track of everything, there would be a major train wreck within seconds.

And I do not at all resent any of the parts for going wrong. On the contrary, having seen what they are up against, I have more respect for them than I had before. I tip my hat to all of them, and I'm glad I'm here outside, wherever that is.

# 21
# SCABIES, SCRAPIE

My wife said: "How can you possibly spend four hours talking about scabies?" This was just before I got into a London cab on my way to a seminar in 1981. Not scabies, I said, *scrapie*, but then the door closed and I had lost the chance to explain scrapie and the remarkable properties of the slow viruses of brain disease which infect today and may not produce symptoms of brain damage until twenty-five years from now, and the overwhelming interest everyone, including my wife, ought to have in forms of life that can replicate themselves without, so far as has yet been discovered, any DNA or RNA at all, one of the greatest puzzles in biology today. Halfway along on the London streets it occurred to me that maybe the virus of scrapie is simply a switched-on, normal gene, and the presumably protein agent that causes the disease may not be alive after all, only a signal to switch on a gene in brain cells which is supposed to be kept off. The notion swamped my mind for most of the cab ride, still does, and I couldn't wait until I got home to explain all this to my wife.

But later in the day, Beryl had her inning. She was reading several books at once, as usual: one by a Cambridge University visiting professor from Poland, with whom we had dined several evenings earlier, on the history of law and religion in the ancient East. Did I know that the Black Stone in Mecca long antedated Islam and that it was supposed to have been pure white originally? Did I know about the Four Dogs of Genghis Khan? Would I like to hear about a fourteenth-century political philosopher named Ibn Khaldun? I would. We went on with it, scrapie from my side of the net, Mesopotamian river basins and their cultural centers from hers.

Thinking back over forty years of marriage, I estimate that I have learned more from her, although she has acquired, for better or worse, for richer or poorer, some of my obsessive concerns, including all of Montaigne, lots of Wallace Stevens, some schoolboy Homer. She probably knows more about endotoxin and the Shwartzman reaction than any academic wife in our acquaintance, although I'm afraid it comes up as a lively topic for conversation only from time to time. I, by contrast, have learned all sorts of things I wouldn't have picked up by myself. Beryl not only reads everything, she remembers everything she reads, and likes to tell me about items as she goes along. I would never have started my way through Jane Austen without this subtle guiding and goading. At one time I bobbed along in her wake, getting partway but never all the way into Proust. Next month I shall start George Eliot, or sometime soon anyway, having watched her elation across the room, evenings, through one novel after another. It was only when I heard her hooting with laughter in the middle of Michael Frayn that I realized what

I'd been missing. Some places I've not been able to go: Anthony Powell and all those books of conversation, all the Paul Scott, medieval cathedral architecture, any number of the English detective stories she collects like porcelain, Scottish history, French history, English history, David Cecil's *Melbourne* (a favorite of hers), the Restoration poets. I rely on her for knowing what colors go with what other colors, and when a particular shade of rose or green in a curtain or a rug or a painting is a particularly good color. I am not color-blind but color-deaf.

I am a skilled gramophone player, better at Bach, I think, and I like to fill the house with the Beethoven late quartets, late nights, full blast. I went through a time when I couldn't have enough of Bartók and Elliott Carter quartets, played over and over again at high volume. A while back I bought a Sony Walkman, an electronic marvel on which I can play cassettes in loud stereo inaudible beyond my ears, but Beryl is not fond of this instrument, preferring me to make music openly and honestly even when—as I suspect—she might like the silence better.

Living together has been like an extended, engrossing, educational game. We have been exchanging bits of information, tastes, preferences, insights for so long a time that our minds seem to work together. My firm impression is that I've come out ahead so far, in the sense that I've been taught more surprising things by her than I've ever stored up to teach in return. But, even asymmetrically, it has continued to work both ways.

On balance, I believe this kind of family education is something women are better at than men. It is surely better accomplished by women for the children of a family. All the old stories, the myths, the poems comprehended most acutely by young children, the poking and nudging and pinching of very young minds, the waking up of very small children, the learning what smiles and laughter are all about, the vast pleasure of explanation, are by and large the gifts of women to civilization. It is the women who remember and pass along the solid underpinnings of culture, not usually the men. The men may come in later with their everyday practical knowledge of what they see as the great world, sometimes with the necessary ambiguities and abstractions, but they cannot fit their contributions into expanding, exploding young minds unless the women—the mothers and wives, the aunts and grandmothers, the elder sisters—have done their work first. I'm not even sure that a child can discover what fun is, and how to have it, without first being led by the hand into fun by a woman.

While I am at it, I might as well tell the whole truth about the sexes, giving away the whole game. It is my belief, based partly on personal experience but partly also arrived at by looking around at others, that childhood lasts considerably longer in the males of our species than in the females. There is somewhere a deep center of immaturity built into the male brain, always needing steadying and redirection, designed to be reconstructed and instructed, perhaps analogous to the left-brain center for male birdsong, which goes to pieces seasonally and requires the reassembling of neurones to function properly when spring comes. Women keep changing the upper, outer parts of their minds all the time, like shifting the furniture or changing their handbags, but the center tends to hold as a steadier, more solid place.

I am, in short, swept off my feet by women, and I do not think they have yet been assigned the place in the world's affairs that they are biologically made for. Somewhere in that other X chromosome are coils of nucleic acid containing information for a qualitatively different sort of behavior from the instructions in an average Y chromosome. The difference is there, I think, for the long-term needs of the species, and it has something to do with spotting things of great importance. To be sure, most women tend to fret more than most men over the small details of life and the rules of behavior, they tend to worry more about how things look, they are more afflicted by the fear of missing trains or losing one glove, they cry more readily. But on the very big matters, the times requiring exactly the right hunch, the occasions when the survival of human beings is in question, I would trust that X chromosome and worry about the Y.

This brings me to a proposal. Taking all in all, the history of human governments suggests to me that the men of the earth have had a long enough run at running things; their record of folly is now so detailed and documented as to make anyone fear the future in their hands. It is time for a change. Put the women in charge, I say. Let us go for a century without men voting, with women's suffrage as the only suffrage. Try it out, anyway. Write into the law, if you like, a provision that men can begin voting again after a hundred years, if they still want to at that time, but in the meantime, place the single greatest issue in the brief span of human existence, the question whether to use or get rid of thermonuclear weapons of war, squarely in the laps of the world's women. I haven't any doubt at all what they will do with this issue, possessing as they do some extra genes for understanding and appreciating children. I do not trust men in this matter. If it is left in their charge, someone, somewhere, answering some crazy signal from a Y chromosome, will start them going off, and we will be done as a species.

Also, another matter. The world has become crowded with knowledge, and there is more to come. The fair redistribution of knowledge will be a more important problem in the century ahead than at any time in the past. Women have not had much hand in this up to now. The full education of children, up through adolescence into early adult life, will soon become the great challenge for humanity, once we have become free of the threat of bombs. All the more reason, I should think, to put the women, born teachers all of them, in charge. Send the men, for the time being anyway (the time being a hundred years), off to the showers, for the long, long bath they have earned.

# 22
# ESSAYS AND GAIA

In 1970, a symposium on "Inflammation" was held at Brook Lodge, somewhere outside Kalamazoo, sponsored by the Upjohn Company, which maintains the place as a conference center for meetings of university scientists. My assignment was to provide something called the keynote address, to precede about forty other papers by researchers, from all over, who were working on the phenomenon of inflammation. Since I had no way of knowing in advance what the papers would be about, I was free to say whatever I liked about the matter. I knew that the general drift of the papers would concern biochemical details of the defensive machinery in living tissues: the ways in which signals are exchanged between leukocytes and other cells in the presence of foreign invaders, mediators governing expansion and constriction of small blood vessels to regulate the passage of blood cells and various soluble components across the walls of capillaries and small veins, the clumping and sticking of leukocytes and platelets inside the vessels, all culminating in the Galenic signs of *rubor* (redness), *calor* (heat), *tumor* (swelling), and *dolor* (pain), the classical inflammatory reaction. This complicated chain of linked events, leading to the destruction and ejection of foreign material lodged in living tissue, was the topic at hand.

This kind of conference tends to be rather heavy going, and my talk was designed to lighten the proceedings at the outset by presenting a rather skewed view of inflammation. I thought of it, and still do, as an example of largely self-induced disease rather than pure defense, with all sorts of mutually incompatible, combative mechanisms turned loose at once, frequently resulting in more damage to the host than to the invader, a biological accident analogous to a multicar accident involving fire engines, ambulances, police cars, and tow trucks all colliding on a bridge.

The whole conference was recorded on tape by the sponsors, and several months later I received in the mail a pamphlet-sized reproduction of my talk, accompanied by a note saying that it was being sent round to the other participants. A day or so later I had a telephone call from Franz Ingelfinger, the editor of the *New England Journal of Medicine*.

Ingelfinger said he had read the piece and liked it, parts of it anyway, although he didn't agree with all of it, and he wanted me to try writing some essays for the *Journal* in the same general style. The terms were attractive enough: I would have to write one essay each month, due on Thursday of the third week, no longer than the space of one *Journal* page (around a thousand words), on any topic I liked. There would be no pay, but in return he would promise that nobody would be allowed to edit an essay. They would print them or not, but not change them.

I could not say no. Not because of the *Journal*'s prestige or the opportunity to publish any idea I liked, but because I had been conditioned, long ago, to doing whatever Ingelfinger told me to do. Partly, this was a behaviorist reflex established by the contingencies of the Boston City Hospital internship. I had come to the service as the pup when Ingelfinger, who had graduated from Harvard a year earlier, was the Senior, nine months ahead of me and, therefore, my boss. Our relationship began, and continued, with him giving the orders and me carrying them out. But there was more to it. Going through the City Hospital internship in those days was something like combat on a disordered battlefield, and we became close friends. Ingelfinger was, among other things, a born teacher, and he set about teaching me everything he knew, from the moment I turned up on his ward in my new white suit. There were a great many small skills to be learned—how to put together a makeshift oxygen tent when the proper parts were missing, how to bring a new oxygen tank to a patient's bedside from the outer corridor where the things were stored (they were much too heavy to carry, and what you did was to tilt them over at exactly the right angle and then run down the ward behind them, sliding them along the waxed floor), how to wash out the stomach of someone in a coma, how to get a needle into an invisible vein, and so forth. Evenings we would sit around his room or mine, waiting for calls from the emergency room, playing records. Ingelfinger knew more about Mozart than I did, and couldn't stop teaching even then: he loved to put a record on, play it for a second or so, then lift the needle and challenge me to identify the phrase and its location. I was on duty Christmas Eve in 1937, and Ingelfinger was off, due in the next morning at seven. It was a quiet night on the wards. I tacked a note on his door to greet him, a Christmas card:

> Of Christmas joy I am the bringer.
> I bring good news to Ingelfinger.
> Though many turned in bed and cried,
> Nobody died, nobody died.

We had eighteen months together, and I came away with a deep respect for his mind and character. After the internship, he went to Philadelphia, I to New York, and in the years that followed we met each other infrequently, once or twice a year, usually at the May meetings of the Society for Clinical Investigation in Atlantic City. Whenever we did meet we settled down to continue last year's conversation wherever we'd left off.

So, I told Ingelfinger I'd be glad to try writing the column for his *Journal*, and began.

I had not written anything for fun since medical school and a couple of years thereafter, except for occasional light verse and once in a while a serious but not very clear or very good poem. Good bad verse was what I was pretty good at. The only other writing I'd done was scientific papers, around two hundred of them, composed in the relentlessly flat style required for absolute unambiguity in every word, hideous language as I read it today. The chance to break free of that kind of prose, and to try the essay form, raised my spirits, but at the same time worried me. I tried outlining some ideas for essays, making lists of items I'd like to cover in each piece, organizing my thoughts in orderly sequences, and wrote several dreadful essays which I could not bring myself to reread, and decided to give up being orderly. I changed the method to no method at all, picked out some suitable times late at night, usually on the weekend two days after I'd already passed the deadline, and wrote without outline or planning in advance, as fast as I could. This worked better, or at least was more fun, and I was able to get started. I finished an essay called "The Lives of a Cell," then one about the precautions against moon germs at the time of the first moon landing, then several about the phenomenon of symbiosis, and after six months I'd had six essays published and thought that was enough. I wrote a letter to Ingelfinger, suggesting that now it was probably time to stop—six essays seemed more of a series than I'd planned, and perhaps the *Journal* would do well to drop the venture and start something new with someone else doing the writing. I got a letter back saying no, I had to keep it up, they were getting letters from readers expressing interest, and in case I had any doubts myself, I should know that even Lowell had telephoned Ingelfinger and said that the Thomas essays were not bad. Lowell was Dr. Francis Cabot Lowell, an intellectually austere Boston classmate of Ingelfinger's and a severe critic. If Lowell approved of the pieces I should surely keep them going.

After a while it became a kind of habit, and I continued writing with fair regularity for something over four years. One day I had a letter from Windsor, Canada, from Joyce Carol Oates, whom I had never met, telling me that a physician had shown her reprints of some of the essays; she advised me to think about collecting them for a book. I had received quite a lot of interesting mail about the essays, mostly from doctors and medical students, but the Oates letter was, and remains, the nicest letter in my files. I didn't see how the pieces could ever be made into a book, since they seemed to me quite unconnected with each other, but I was encouraged by her approval. A bit later, in 1973, I had notes from several publishing houses inquiring about the possibility of a book, but also letting me know that if I was interested in trying I would have to do a lot of rewriting and insert some new essays—"connecting pieces," they were called—in order for the thing to make sense. I was preoccupied in the dean's office at Yale, and wrote back that I couldn't manage the work. Then Elisabeth Sifton, an editor at The Viking Press, telephoned one morning to say that she would like to publish the essays as they were, no patching or connecting for me to do, and I said yes over the phone. They came out in 1974, titled for the first essay, *The Lives of a Cell.* It was the easiest of books to write, and I was surprised that it did well in the marketplace, especially pleased that its most active market was in university and medical school bookstores, which was what I'd hoped for. Having caught the habit, I kept on writing short essays, some for the *New England Journal*, some unpublished, and four years later there were enough for a second book, *The Medusa and the Snail.*

Although the books were favorably reviewed in *Nature* and *Science*, and a good many basic scientists have written letters of general approval and support, I have gotten myself in trouble with several groups of very knowledgeable people who are closely involved in subjects that I've written about and who disagree with me vigorously, not so much on matters of fact as on points of view.

Some of the evolutionary biologists criticize the suggestion, running through many of the essays, that the earth's body represents a kind of organism, displaying so many instances of interdependency and connectedness as to resemble an enormous embryo still in the process of developing. This notion has seemed reasonable enough to me, considering the plain paleontological fact that the life of the earth was, for nearly 75 percent of its existence, made up entirely of separate, prokaryotic, microbial cells, themselves the progeny of what may have been, long ago, the single first cell, and these somehow succeeded in developing into nucleated cells around a billion years ago and then into multicellular entities, culminating in today's elaborate plants and animals, all of which live their lives in dependence, direct or indirect, on today's microbial populations. The Gaia hypothesis, proposed by Lovelock and supported by Margulis, goes a step further to postulate that the conjoined life of the planet not only comprises a sort of organism but succeeds in regulating itself, maintaining stability in the relative composition of the constituents in its atmosphere and waters, achieving something like the homeostasis familiar to students of conventional complex organisms, man himself for example.

The concentrations of oxygen, carbon dioxide, and nitrogen in the atmosphere are optimal for life, and they are wildly different from what would be predicted for the atmosphere of a lifeless planet. There is substantial evidence that each of these, and other trace gases such as ammonia and methane, are held at *optimal* concentrations by the metabolic activity of various forms of life. The salinity and pH of the oceans remains constant, in spite of trends which could have been expected to turn the seas to brine long since if it were not for life. The earth's *mean* temperature is fixed somewhere between 10 degrees and 20 degrees Centigrade, and has been this way for at least a billion years. The life of the planet maintains the life, in the face of forces that would, in the absence of an intricate mechanism for keeping the balance of things constant in the environment, have led inevitably to death everywhere. Moreover, and most spectacularly of all, the biosphere has managed to adjust, and to keep on adjusting, to a steady increase in the luminosity of the sun amounting to at least 30 percent since the beginning of life 3.5 or so billion years ago. The place is remarkably stable. It begins to look as though the only real threat to its abundant continuance may be us.

The arguable trouble with the Gaia idea is that it seems to violate the central doctrine of Darwinian theory, that species emerge and then survive or fail on grounds of competition for natural selection, and that the appearance of any new species in candidacy for survival is a matter of pure chance, governed by random mutations or by equally random rearrangements of existing genes. According to some of the evolutionary biologists, the suggestion that the whole apparatus, Gaia, now possesses an array of unfathomably complex mechanisms for its own internal regulation, and that these mechanisms, operating together, maintain the stability of the whole, carries the implication that the thing was somehow designed to work this way. This would be against the rules, for as one critic said of the idea, "Natural selection cannot plan ahead." I agree with this, but I do not regard the notion as all that teleological, and certainly not, as the same critic asserted, a "deification" of the earth. It seems to me that there are solid biological advantages in behavior that results in cooperation and collaboration, and the record of the earth abounds in successful examples of partnership, starting early on in the algal mats in which much of the earth's life was embedded, in mutually dependent layers, more than a billion years ago, perhaps culminating in the entry of microorganisms into the flesh of other microorganisms to form the chloroplasts and mitochondria. This I view as the most fundamental of all stages, making possible everything that came on the scene in the last billion years: a system of increasingly complicated metazoan life, plants and animals, equipped to utilize the sun's energy for the making of food and oxygen, and, in turn, using the oxygen to gain the energy stored in the food. I have no doubt that it happened by chance and succeeded by natural selection, but it happened with such dazzling success that it set the stage for all sorts of later events which would not have occurred without it. These primary, primal endosymbionts have retained enough of their own DNA and RNA to inform the molecular geneticists that they are, or were, indeed, separate living things to begin with, but they have probably turned over some of the essential genes for their structure and function to the nuclei of the cells in which they now live.

I confess that I am overwhelmed by what I take to be the meaning of this piece of natural history, and perhaps because of this I tend to look for as many other examples of symbiotic living as I can find in the record. Perhaps also, with my mind set in this bias, I pay less attention than I should to the examples of aggression, grabbiness, and terrorism in nature. I cannot help this. I suppose I could change my mind and spend my time thinking about insects as mainly parasites, mainly spreaders of disease, mainly an affliction, instead of thinking of them as the major, indispensable sources of food for aerial and aquatic life, carriers of heredity for plants, and technical assistants in the recycling of forest life. It is a Panglossian bias, I acknowledge, but I'm not sure Pangloss was all that wrongheaded. This is in real life the best of all *possible* worlds, provided you give italics to that word possible.

I am not so optimistic about us. Human beings are getting themselves, and the rest of the world, into deeper and deeper trouble, and I would not lay heavy odds on our survival unless we begin maturing soon. Up to now, we have been living through the equivalent of an early childhood for our species. We have not been here any length of time, in evolutionary terms, and no wonder we are still young, with nothing but frontal lobes, thumbs, language, and culture to rely on for our shelter and survival. We could fumble and do it wrong. Thermonuclear war is the worst case to contemplate, enough in itself to cause the crash of the species, but we have other threats to make against our lasting existence: overpopulation and crash, deforestation and crash, pollution and crash, a long list of possible bad dreams come true, the sounds, always outside the window offstage, of the destruction of the orchard. With luck we may come through. The luck will have to be incalculable, and unbelievably on our side over the next few decades. The good thought I have about this is that we are, to begin with, the most improbable of all the earth's creatures, and maybe it is not beyond hope that we are also endowed with improbable luck.

# APPENDIX

What follows is a series of notes and references which are in no sense comprehensive or scholarly. They do not form a real bibliography, since most of the references are to studies from my own laboratory and only a few are to the body of much more important work on the same or related matters by others. I have keyed them by page number in order to provide convenient sources for those who may be curious about some of the details. Here and there, I have also included comments that could not be fitted neatly into the chapters themselves. And I have used the whole section as an excuse to insert some items of verse that I grew fond of long ago when I wrote them. Anyone who wishes to skip the whole Appendix will be missing very little.

p. 5

Allen Street was what everyone called the morgue at the Massachusetts General Hospital: the morgue doorway opened on that street. Herewith the poem.

## ALLEN STREET

### *Canto I: Prelude*

Oh, Beacon Street is wide and neat, and open to the sky—
Commonwealth exudes good health, and never knows a
sigh—
Scollay Square, that lecher's snare, is noisy but alive—
While sin and domesticity are blended on Park Drive—
And he who toils on Boylston Street will have another day
To pay his lease and live in peace, along the Riverway—
A thoroughfare without a care is Cambridge Avenue,
Where ladies fair let down their hair, for passers-by to
view—
Some things are done on Huntington, no sailor would
deny,
Which can't be done on battleships, no matter how you
try—
Oh, many, many roads there are, that leap into the mind
(Like Sumner Tunnel, that monstrous funnel, impossible to
find!)
And all are strange to ponder on, and beautiful to know,
And all are filled with living folk, who eat and breathe and
grow.

### *Canto II*

But let us speak of Allen Street—that strangest, darkest
turn,
Which squats behind a hospital, mysterious and stern.
It lies within a silent place, with open arms it waits
For patients who aren't leaving through the customary
gates.
It concentrates on end results, and caters to the guest
Who's battled long with his disease, and come out second-
best.

For in a well-run hospital, there's no such thing as death.
There may be stoppage of the heart, and absence of the
breath—

But no one dies! No patient tries this disrespectful feat.
He simply sighs, rolls up his eyes, and goes to Allen Street.
Whatever be his ailment—whate'er his sickness be,
From "Too, too, too much insulin" to "What's this in his
pee?"
From "Gastric growth," "One lung (or both)," or
"Question of Cirrhosis"
To "Exitus undiagnosed," or "Generalized Necrosis"—
He hides his head and leaves his bed, and, covered with a
sheet,
He rolls through doors, down corridors, and goes to Allen
Street.

And there he'll find a refuge kind, a quiet sanctuary,
For Allen Street's that final treat—the local
mortuary.

*Canto III*

Oh, where is Mr. Murphy with his diabetic ulcer,
His orange-red precipitate and coronary?
Well, sir,
He's gone to Allen Street.
And how is Mr. Gumbo with his touch of acid-fast,
His positive Babinskis, and his dark luetic past?
And what about that lady who was lying in Bed 3,
Recently subjected to such skillful surgery?
And where are all the patients with the paroxysmal
wheezes?
The tarry stools, ascitic pools, the livers like valises?
The jaundiced eyes, the fevered cries, and other nice
diseases:
Go! Speak to them in soothing tones. We'll put them on
their feet!
We'll try some other method, some newer way to treat—
We'll try colloidal manganese, a diathermy seat,
And intravenous buttermilk is very hard to beat—
We'll try a dye, a yellow dye, or different kinds of heat—
But get them on their feet—

We'll find some way to treat—
I'm very sorry, Doctor, but they've gone to
Allen Street. . . .

*Canto IV*

Little Mr. Gricco, lying on Ward E,
Used to have a rectum, just like you or me—
Used to have a sphincter, ringed with little piles,
Used to sit at morning stool, face bewreathed with smiles,
Used to fold his *Transcript*, wait in happy hush
For that minor ecstasy, the peristaltic rush. . . .
But in the night, far out of sight, within
his rectal stroma,
There grew a little nodule, a nasty
carcinoma.
Oh, what lacks Mr. Gricco?—Why looks he incomplete?
What is this aching, yawning void in Mr. Gricco's seat?
Who made this excavation? Who did this foulest deed?
Who dug this pit in which would fit a small velocipede?
What enterprising surgeon, with sterile spade and trowel,
Has seen some fault and made assault on Mr. Gricco's
bowel?
And what's this small repulsive hole, which whistles like a
flute?
Could this thing be colostomy—this shabby substitute?
Where is this patient's other half! Where is this patient's
seat!
Why, Doctor, don't you recollect: It's
gone to Allen Street.

*Canto V: Footnote*

At certain times one sometimes finds a patient in his bed,
Who limply lies with glassy eyes receding in his head.
Who doesn't seem to breathe at all, who doesn't make a
sound,

Whose temperature is seen to fall, whose pulse cannot be
found.
And one would say, without delay, that this is a condition
Of general inactivity—a sort of inanition—
A quiet stage, a final page, a dream within the making,
A silence deep, an empty sleep without the fear of
waking—

But no one states, or intimates, that maybe he's expired,
For anyone can plainly see that he is simply tired.
It isn't wise to analyze, to seek an explanation,
For this is just a new disease, of infinite duration.

But if you look within the book, upon his progress sheet,
You'll find a sign within a line—"Discharged to Allen
Street."

*p. 5*

Considering everything (inflation, taxes, etc.), you can match the incomes of 1937 to those of today, very roughly, by multiplying by five. Thus, the twenty-year graduates were making, in today's money, approximately $37,500. It seems safe to say that the average Harvard Medical School alumnus of the class of, say, 1963, makes a lot more than that, by two or three times, considerably more if engaged in a surgical specialty.

The cost of a medical education, however, was very much less. Tuition, room, and board, in the years between 1922 and 1937, came to around $650. Today, more than twenty times that sum would be needed for the minimum cost of a year in Harvard Medical School. Most students, or their families, must take out loans (formerly sponsored by the federal government at low interest but now costing the moon), and it is common for students to graduate under a high-interest debt of $60,000 or more.

*p. 23*

My mother and father were married on October 31, 1906. The only record I possess of this event is a page in the center of the family

Bible, which still rests for most of its time on a back shelf of a closet, too big a volume for housing in an ordinary bookcase. Someday, I tell myself, I must bring that page up to date. As it stands, it ends with a note in my father's handwriting recording my marriage to Beryl and the birth dates of our three daughters, followed by similar notes about my brother Joe's marriage some years later. There is a lot more to be filled in, and I must do it soon.

*p. 32*

The discovery that liver extract was a cure for pernicious anemia was made in 1926. The bare outline of the story is as follows: Dr. George Whipple, at the University of Rochester, reported in 1925 that dogs that had been made anemic by repeated bleedings were helped in reconstituting their blood by being fed large quantities of fresh liver. Because of this, Dr. George Minot decided to try feeding liver to patients with pernicious anemia and, aided by his younger associate Dr. William P. Murphy at the Peter Bent Brigham Hospital, quickly observed that the disease was indeed cured. Years later, others found that the active ingredient in liver extract responsible for the cure was a complex cobalt-containing molecule now known as vitamin $B_{12}$.

That was the essence of the story, and probably all that is needed to illustrate the neatness and precision with which biomedical research can sometimes solve major disease problems. Until the Minot-Murphy-Whipple discovery, pernicious anemia had been a universally fatal disease and a blank mystery.

But the story is more complicated than this outline. A somewhat longer version was told to me by one of Minot's colleagues, who had been around on the wards of the Peter Bent Brigham at the time. If it is true, which I cannot prove, it illustrates another important aspect of medical discovery—the role of luck.

According to this colleague, Minot customarily recruited junior associates for his busy private practice in hematology from the ranks of residents at Peter Bent Brigham, and Murphy was next in line for such a job. Minot met with Murphy to discuss the position, and advised the younger man to take on a research project of some kind for the next few months in his spare time; Minot's office associates were expected to have at least one publishable paper at the time of starting practice. Minot leafed through a stack of journals on his desk and came across Whipple's paper on anemia in dogs. Why not try the same sort of liver extract in patients with anemia? What kind of anemia? Well, why not pernicious anemia? There were always some patients with this disease on the Peter Bent Brigham wards.

So much for that part of the luck. There was more, if my story is true. The earliest versions of raw liver prepared for feeding were extremely unpalatable, and the first patient selected for trial of the therapy was an exceedingly resistant elderly woman whose personality had been so affected by her illness that she had the reputation of being obstreperous, cantankerous, and impossible to deal with. Murphy regarded her as a double challenge to his ability: to test the treatment and to persuade her to cooperate. It was a contest of wills, and Murphy, through sheer stubbornness, won. On about the seventh day of battle, her personality abruptly changed, she was transformed into an agreeable, accommodating woman, and a few days later her blood began to show the now-familiar signs of recovery from pernicious anemia.

And now the final piece of luck. Looking back at the events, doctors are now generally agreed that Whipple's dogs could have had nothing at all like pernicious anemia. Their anemia was actually due to iron deficiency brought about by repeated hemorrhage, and the response to liver was almost undoubtedly caused by the iron contained in the very large doses of liver used for feeding. It was the wrong model to use for studying pernicious anemia, and it led straight to a Nobel prize.

Minot was personally lucky in still another sense. He became ill with severe diabetes in 1921 and by 1922 was being maintained on the near-incapacitating dietary restriction that was the only treatment available at the time. His weight had dropped to 120 pounds (he was 6 feet tall). Had it not been for the Banting-Best discovery of insulin in 1922, it is unlikely that Minot would have lived to find the cure for pernicious anemia.

I knew a New York internist, a professor of medicine at P&S,

who claimed that he could have made Minot's discovery and gotten the Nobel prize himself if he'd had his wits about him in the early 1920s. He had a patient with all the blood manifestations of pernicious anemia, a wealthy spinster living in an apartment on Park Avenue, who baffled him and his hematologist colleagues by remaining in vigorous good health despite her disease. He should have looked into her kitchen, or inquired more closely into her diet. Years later, after the Minot discovery was announced, she told him that she had always adored pâté de foie gras, and ate jar after jar all through those years.

*p. 37*

I discovered, by accident, a new way of adding to my income during my internship and residency. I had been writing verse off and on, late nights, while on call, using the secretary's typewriter in the Harvard office on the top floor of the Peabody. One night I forgot to bring the folder back to my room, and Dr. A. P. Meiklejohn, a Scottish physician then working in the nutrition laboratory at the Thorndike, found it the next morning. Without telling me, he sent copies to friends of his in New York who were running a literary agency, Russell (A.E.'s son) and Volkening. A few weeks later I had a letter from Russell saying they'd be glad to try selling some of the poems. *The Atlantic Monthly*, *Harper's Bazaar*, and *The Saturday Evening Post* bought about a dozen items all together, at $35 each, and I had a steady income. With two transfusions and one poem sold each month, I lived quite well.

When the first poem turned up in *The Atlantic*, I received a handwritten note from Dr. Minot in his official capacity as chief of the Harvard Medical Services at the Boston City.

Dear Thomas—

I knew you were an excellent physician and student of medicine, but not until this evening did I know you were a poet. I have just seen the December issue of *The Atlantic*

*Monthly.* I consider it a splendid achievement to "make" *The Atlantic Monthly.* I wish I had known of this earlier today when I saw Mr. Weeks, the editor.

This note carries my sincere congratulations to you.

As ever yours,
George R. Minot

I still have these old magazines, on a back shelf in my closet, along with my old violin, which I keep telling myself I will exhume someday, restring, and play again—but haven't, for forty years. I've forgotten exactly when the poems were written, but all of them came from the time between 1938 and Pearl Harbor. The atom bomb was nowhere in sight, but even so the new technologies of warfare seemed to have no limits.

### MILLENNIUM

It will be soft, the sound that we will hear
When we have reached the end of time and light.
A quiet, final noise within the air
Before we are returned into the night.

A sound for each to recognize and fear
In one enormous moment, as he grieves—
A sound of rustling, dry and very near,
A sudden fluttering of all the leaves.

It will be heard in all the open air
Above the fading rumble of the guns,
And we shall stand uneasily and stare,
The finally forsaken, lonely ones.

From all the distant secret places then
A little breeze will shift across the sky,
When all the earth at last is free of men
And settles with a vast and easy sigh.

*p. 37*

When I was five months along in my internship, the hierarchy in command of the wards consisted, in rank order, of Dr. William Peltz, House Physician; Cary Peters, Assistant House; and Franz Ingelfinger, Senior Physician. The names of all three are embedded in a poem which I have kept in a folder of unpublished verse from those years. This one concerns a patient named A. Maloof who came to my ward terribly ill with some disease that I've now forgotten. I can't explain why I've forgotten; you might expect an intern from that time to remember anyone who came in terribly ill and then got well, but anyway I've forgotten. What I do remember is the gift he brought in later as a token of gratitude to all his doctors. It was a sort of antique jug, about two feet tall, solid brass, with the strangest-looking spout I'd ever seen on a pot. Since Peltz was the House Physician, he claimed rights to the object. There was no argument, but I thought the event needed memorializing.

### ODE TO MALOOF

Oh, Gift of Brass! Oh, splendid proof
Of gratitude from A. Maloof—
                —If thou could'st speak!
What tales of years fourscore and ten,
What memories of now and then
                Would pass that beak?
What hopes had thou? Did e'er thy heart
Aspire to be an *objêt d'art?*
                Despite the beak?
Who fashioned thee? What mad designer
Did pound out brass in Asia Minor
                And made that beak?
What idiot child, with halt and stammer,
Was given brass, and handed hammer,
                And forced to stay away, aloof
                From others of the clan Maloof,

To use his tortured, fevered mind,
To make this pot on which we find
This Monstrous Beak?
And did'st thou hope, oh, Burnished Pot,
That such a fate would be thy lot?—
That thou would'st be so doubly blest:
To leave Maloof, and be the guest
Of someone else?
Of William Peltz?
Oh, Happy Pot! Oh, lucky toss!
Maloof came in while Peltz was boss!
For if he'd chose at home to tarry,
Thou might'st have gone to Peters,
Cary!
And if he'd longer chose to linger,
Thou would'st have gone to
Ingelfinger!
But No! Thou goest to no one else
But Dr. William Learned Peltz!

*p. 47*

The effects of alcohol on the human brain were a matter of lively interest in the Boston City Hospital because of the great numbers of patients with advanced alcoholism on most of the wards. Professor Leo Alexander, an Austrian neuropathologist, worked in the late 1930s on Wernicke's disease, a strange affliction involving the brain centers for eye movement, which he believed was due to the combined effects of alcohol and nutritional deficiency. He had some experiments going on in his laboratory with pigeons, in which he claimed that brain lesions similar to those of Wernicke's disease could be produced. He announced these results at a seminar, followed by a celebratory dinner. A friend of mine, Joseph Ross, working in the pathology department, was invited to make a few remarks at the dinner and asked if I could write a piece of light verse for him to read. I had just given a transfusion that afternoon and, with the help of my allotment of Golden Wedding, produced the following lines. I should add that Professor Alexander was an impressive figure, very dignified, portly, unmistakably Viennese; each spring he shaved his large head bald.

LINES DEDICATED TO DR. LEO ALEXANDER

Hail to Alexander, that giant Man of Science,
And all the happy pigeons, his little waddling clients,
A happy lot, I truly wot, to meditate upon—
Each little bird, with vision blurred, may daily tie
one on.

May daily have his highball, with little ice cubes
clinking,
May walk around his cage all day while absolutely
stinking,
May peer upon the moving world with slight
nystagmus trouble,
May know the rich experience of seeing Leo double!

But now remove his sustenance, deprive him of his B,
Withdraw his food and let him brood on crusts and
Irish tea.
—He undergoes Mutation, becomes like me or you,
Becomes an almost-human thing, and ends up on
Ward 2.

A pigeon with religion, a pigeon with the shakes,
A little dove protesting love to twenty thousand
snakes,
A pigeon having horrors of being hung on hooks,
Or being chased and then defaced by busts of Phillips
Brooks.
Oh, Happy Lot! Oh, Joyous News! Oh, Science on
the Brink!
The Mammillary Bodies are susceptible to drink!

The Region of the Thalamus delights in getting
blotto,
And also the entire damned Medulla Oblongato!

*p. 73*

Rivers, T. M. and Schwentker, F. F. Encephalomyelitis accompanied by myelin destruction experimentally produced in monkeys. *Journal of Experimental Medicine*, 61: 689, 1935.

*p. 74*

Thomas, L. A single-stage method to produce brain abscess in cats. *Archives of Pathology*, 33:472–76, 1942.

*p. 75*

Dingle, J. H., Thomas, L. and Morton, A. R. Treatment of meningococcic meningitis and meningococcemia with sulfaldiazine. *Journal of the American Medical Association*, 116: 2666–68, 1941.

*p. 78*

Thomas, L., Smith, H. W. and Dingle, J. H. Investigations of meningococceal infection. *Journal of Clinical Investigation*, 22: 353–85, 1943.

*p. 101*

Thomas, L. and Peck, J. H. Results of inoculating Okinawan horses with virus of Japanese B encephalitis. *Proceedings of the Society for Experimental Biology and Medicine*, 61: 5–6, 1946.

*p. 103*

Alvin Coburn must have been one of the country's all-time most successful interns. He graduated from P&S with a consuming interest in rheumatic fever and some strong hunches about its causation, and during the months of his internship, in the midst of all the endless duties of a house officer, somehow or other he managed to solve the most crucial part of the problem. Single-handed, in what there was of his spare time, he established the causal linkage between infection of the throat by Group A hemolytic streptococci and the onset, ten or twelve days later, of inflammation affecting the heart and joints.

Not that he wasn't working on his wards. A co-intern of his told me that Coburn would arrive on his ward promptly at 6:45 each morning, stand in the doorway, flap his arms, and crow like a rooster.

I'll never know how he accomplished that masterpiece of applied microbiology so clearly and conclusively and in so short a space of time. So far as I know, although he continued to study rheumatic fever for most of the rest of his professional life, he never matched that single achievement, and he never received all the credit he deserved.

*p. 111*

Avery, O. T., MacLeod, C. M. and McCarty, M. Transformation of pneumococcal types induced by a desoxyribonucleic acid fraction isolated from Pneumococcus Type III. *Journal of Experimental Medicine, 79:* 137, 1944.

I have never understood why this work, one of the great pieces of biological science in history, was never recognized by a Nobel prize. The paper in which it was first announced that the "transforming principle" of pneumococci was DNA was written with meticulous care and caution but in full awareness of the implications for genetics. Avery and his associates knew exactly what they had, and what it meant. Some have said that the writing was *too* cautious, that perhaps Avery had doubts of his own, and this is why the prize was withheld. I cannot believe it.

*p. 147*

Simon's original notion was that there might be specific sets of cells within the nervous system to which morphine and related opiates are specifically bound. He obtained suggestive evidence for this in 1966, and published a preliminary note in the *Proceedings of the Society for Experimental Biology and Medicine* (122: 6–11, 1966). Seven years later conclusive proof of the phenomenon was obtained by his group and by Pert and Snyder at Johns Hopkins. Within two years, in 1975, other investigators discovered the existence of endogenous opiates (the "endorphins") in pituitary gland secretion. The history of this remarkable work, involving several different laboratories here and abroad, can be found in two review articles published in 1978.

Terenius, L. Endogenous peptides and analgesia. *Annual Review of Pharmacology and Toxicology*, 18: 189–204, 1978.

Simon, E. J. The opiate receptors. *Annual Review of Pharmacology and Toxicology*, 18: 371–94, 1978.

*p. 150*

Thomas, L. The physiological disturbances produced by endotoxins. *Annual Review of Physiology*, 16: 467, 1954.

*p. 155*

Good, R. A. and Thomas, L. Studies on the generalized Shwartzman reaction. *Journal of Experimental Medicine*, 97: 871–88, 1953.

*p. 157*

Thomas, L. Reversible collapse of rabbit ears after intravenous papain and prevention of recovery by cortisone. *Journal of Experimental Medicine*, 104: 245–52, 1956.

*p. 158*

Barber, B. and Fox, R. The case of the floppy-eared rabbits: An instance of serendipity gained and serendipity lost. *American Journal of Sociology,* 64: 128–36, 1958.

*p. 159*

Thomas, L., Douglas, G. W. and Carr, M. C. The continual migration of syncytial trophoblasts from the fetal placenta in the maternal circulation. *Transactions of the Association of American Physicians,* 72: 140–48, 1959.

*p. 164*

Fell, H. B. and Thomas, L. Comparison of the effects of papain and vitamin A on cartilage: the effects on organ cultures of embryonic skeletal muscle. *Journal of Experimental Medicine,* 111: 719–44, 1960.

Thomas, L., McCluskey, R. T., Potter, J. L. and Weissman, G. Comparison of the effects of papain and vitamin A on cartilage: the effects in rabbits. *Journal of Experimental Medicine,* 111: 705–18, 1960.

*p. 165*

Thomas, L. Papain, vitamin A, lysosomes and endotoxin. An essay on useful irrelevancies in the study of tissue damage. *Archives of Internal Medicine,* 110: 782–86, 1962.

*p. 175*

I was appointed by Lyndon Johnson to membership on the President's Science Advisory Committee (known in academic circles as PSAC) in 1967. The committee, composed principally of physicists and chemists at that time, met for two days each month in the

Executive Office Building next to the White House under the chairmanship of the President's Science Adviser (in the years of my term, the advisers were Donald Hornig, then Lee Du Bridge, then Edward David). In 1968 we were asked by the White House to make a study of the scientific needs of the country's health-care system, with special attention to the rapidly rising costs of hospital care. I became chairman of a small panel that worked on the matter for about a year and emerged with the conclusion that the only hope of reducing expenditures for health care lay in better and more fundamental research on the mechanisms of human disease, which meant, in our view, more funds for NIH. We recognized three levels of medical technology: (1) genuine high technology, exemplified by Salk and Sabin poliomyelitis vaccines, which simply eliminated a major disease at very low cost by providing protection against the three strains of virus known to exist; (2) "halfway" technology, applied to the management of disease when the underlying mechanism is not understood and when medicine is obliged to do whatever it can to shore things up and postpone incapacitation and death, at whatever cost, usually very high cost indeed, illustrated by open-heart surgery, coronary artery by-pass, and the replacement of damaged organs by transplanting new ones (at extremely high cost); and (3) nontechnology, the kind of things doctors do when there is nothing at all to be done, as in the care of patients with advanced cancer and senile dementia. We suggested that the rising cost of health care was resulting from efforts to treat diseases of the halfway or nontechnology class, and recommended that more basic research on these ailments be sponsored by NIH.

Our report was approved by the full committee and sent along to the President's office, where I imagine it was filed and probably forgotten. The Vietnam War was becoming the sole preoccupation of the Johnson White House, and medical science was not among the high priorities. A few years later, in the Nixon administration, PSAC itself was abolished, I think because its members, almost all of whom came from the academic world, were regarded as too liberal for that administration. Also, I am sure, because some of the members made public their opposition to the Antiballistic Missile (ABM) program then being contemplated, as well as their disagreement with the administration's plans for sponsoring the development of supersonic aircraft on a grand scale.

Too bad, too. PSAC, the Science Adviser, and the staff of the White House Office of Science and Technology were a potentially valuable resource for the government, and I hope that some future President will put the apparatus together again. The need for highly sophisticated, impartial, and sensible advice on matters of science and technology will surely become more and more urgent in the years ahead, and I do not believe that such advice is sufficiently at hand within the bureaucracy of administrative agencies. I would be feeling considerably less apprehensive about the immediate hazard of nuclear warfare if PSAC, or something like PSAC, were back in place. If it were up to me, I would leave off the medical people and biologists, or perhaps have them there as a small minority, and I would load the committee with the best physicists in the United States.

*p. 180*

Thomas, L. Mechanisms of pathogenesis in mycoplasma infection. *The Harvey Lectures.* New York: Academic Press, 1969, pp. 73–98.

*p. 189*

Sabin, A. and Warren, J. *Journal of Bacteriology,* 40: 828, 1940.

*p. 200*

Milton, G. W. Self-willed death or the bone-pointing syndrome. *Lancet,* 1435–36, 1973.

*p. 204*

Thomas, L. Discussion of P. B. Medawar, in *Cellular & Humoral Aspects of the Hypersensitive States.* H. Sherwood Lawrence, editor. New York: Hoeber/Harper, Publishers, 1959, pp. 451–68.

*p. 210*

McCartney, W. *Olfaction and Odours.* Berlin-New York: Springer-Verlag, 1968.

This book is a rich source of convincing anecdotes about the tracking behavior of trained dogs, with the best bibliography I have seen. McCartney has collected references to a vast amount of European and British literature, much of it written by police investigators, some of it published by dog breeding societies. The most active period of research was between the two world wars. Surprisingly, to me anyway, there has been relatively little research recorded in the last three decades.

*p. 211*

Theodor, J. L. The distinction between "self" and "non-self" in lower invertebrates. *Nature,* 227: 690–92, 1970.

Hildemann, W. H., Bigger, C. H., Johnston, I. S., Jokiel, P. L. Characteristics of transplantation immunity in the sponge. *Transplantation,* 30: 362–67, 1980.

*p. 212*

Thomas, L. Biological signals for self-identification. In *Progress in Immunology Vol. 2,* editors: L. Brent and J. Holborow. Amsterdam: North-Holland Publishing Co., 1974, pp. 239–47.

Thomas, L. Symbiosis as an immunologic problem. In *The Immune System and Infectious Disease.* Basel: S. Kargen, 1975, pp. 2–11.

Thomas, L. *The Lives of a Cell.* New York: The Viking Press, 1974, p. 19.

*p. 213*

Yamazaki, K., Boyse, E. A., Mike, V., Thaler, H. T., Mathiesen, B. J., Abbott, J., Boyse, J., Zayas, Z. A. and Thomas, L. Control of mating preferences in mice by genes in the major histocompatibility complex. *Journal of Experimental Medicine, 144:* 1224–35, 1976.

Boyse, E. A., Yamazaki, K., Yamaguchi, M. and Thomas, L. Sensory communication among mice according to their MHC types. In *The Immune System. Functions and Therapy of Dysfunction,* edited by G. Doria and A. Eshkol. New York: Academic Press, 1980, pp. 45–63.

*p. 215*

Yamaguchi, M., Yamazaki, K., Beauchamp, G. K., Bard, J., Thomas, L., and Boyse, E. A. Distinctive urinary odors governed by the major histocompatibility locus of the mouse. *Proceedings of the National Academy of Sciences, 78:* 5817–20, 1981.

*p. 234*

The scabies-scrapie conversation took place one day in 1981, when we were having our third semisabbatical at Cambridge and were in London for a weekend. I had been invited by Sir Michael Stoker, the president of Clare Hall, to come as a Visiting Fellow at Clare Hall for the Michaelmas term, from September to December. We lived in rooms in the college on Herschel Road, a few hundred yards from the university library, with no responsibilities beyond a few lectures and seminars and time enough to write and read through all the rainy days. Our nearest neighbor was a Visiting Fellow from Poland, Professor Gregory Seidler, Rector of Lublin University, a considerable scholar in the philosophy and history of law, and a vivacious, indefatigable conversationalist. The Solidarity crisis and military takeover in Poland were in the British newspaper headlines throughout the time of our friendship, but Seidler and his wife talked of other things. We had a long lunch one Sunday at the Fire Engine House in Ely, a small restaurant just beyond the shadow of

Ely Cathedral, and Seidler spoke his mind on politics for the first and only time. Poland was part of Europe, he said; he and his wife were Europeans. The political problems in Poland were out of hand, beyond solving, for the time being anyway, because they had become "emotional" problems, a disaster. He hoped that the people of Europe could begin thinking together, using their excellent brains, avoiding "emotion." When he spoke of emotion and the problems it raised for his country, he became red in the face.

He presented Beryl with a copy of the English translation of his book *The Emergence of the Eastern World* (Oxford: Pergamon Press, 1968), his only copy, I fear. I have not read it (no time, I keep saying) but Beryl has, and I have been learning, secondhand, some things I never knew about the influence of the Mongol empire on the development of European civilization. I had never heard of Genghis Khan's mother, but now I know that her name was Yulun and that she was a powerful lady, capable all by herself of persuading the nomadic tribes that they should be unified under the leadership of her son.

Professor Seidler and his wife went back to Poland in early January 1982. We got a letter from them in New York, mailed from Cambridge just before their departure, wishing us well. I have never known a more civilized, equable, deeply skeptical but still hopeful man.

*p. 240*

Thomas, L. *Adaptive Aspects of Inflammation.* Presented at Third Symposium of the International Inflammation Club. Published by Upjohn Co., Kalamazoo, 1970.

*p. 245*

Lovelock, J. E. and Margulis, L. "Atmospheric homeostasis by and for the biosphere: The Gaia hypothesis." *Tellus, 26:* 2, 1973.

*p. 269*

Lovelock, J. E. *Gaia. A New Look at Life on Earth.* New York: Oxford University Press, 1979.

*p. 270*

Margulis, L. *Symbiosis in Cell Evolution. Life and its Environment on the Early Earth.* San Francisco: W. H. Freeman & Co., 1981.